Sound Affects

thinking|media

series editors
bernd herzogenrath
patricia pisters

Sound Affects

A User's Guide

Edited by
Sharon Jane Mee and Luke Robinson

BLOOMSBURY ACADEMIC
NEW YORK • LONDON • OXFORD • NEW DELHI • SYDNEY

BLOOMSBURY ACADEMIC
Bloomsbury Publishing Inc
1385 Broadway, New York, NY 10018, USA
50 Bedford Square, London, WC1B 3DP, UK
29 Earlsfort Terrace, Dublin 2, Ireland

BLOOMSBURY, BLOOMSBURY ACADEMIC and the Diana logo are trademarks
of Bloomsbury Publishing Plc

First published in the United States of America 2023
Paperback edition published 2024

Cover design: Daniel Benneworth-Gray
Cover image © Paolo Sanfilippo

Library of Congress Cataloging-in-Publication Data

Names: Mee, Sharon Jane, editor. | Robinson, Luke, editor.
Title: Sound affects: a user's guide / edited by Sharon Jane Mee and
Luke Robinson.
Description: New York : Bloomsbury Academic 2023. | Series: Thinking media
| Includes bibliographical references and index. | Summary: "A philosophical
analysis of sonically charged concepts to map a theory of "sound affects." "–
Provided by publisher.
Identifiers: LCCN 2022025106 (print) | LCCN 2022025107 (ebook) |
ISBN 9781501388880 (hardback) | ISBN 9781501388927 (paperback) |
ISBN 9781501388897 (epub) | ISBN 9781501388903 (pdf) | ISBN 9781501388910
Subjects: LCSH: Sound (Philosophy) | Sound in motion pictures.
Classification: LCC B105.S59 S68 2023 (print) | LCC B105.S59 (ebook) |
DDC 534/.301–dc23/eng/20220705
LC record available at https://lccn.loc.gov/2022025106
LC ebook record available at https://lccn.loc.gov/2022025107

ISBN: HB: 978-1-5013-8888-0
PB: 978-1-5013-8892-7
ePDF: 978-1-5013-8890-3
eBook: 978-1-5013-8889-7

Series: Thinking Media

Typeset by Deanta Global Publishing Services, Chennai, India

To find out more about our authors and books visit www.bloomsbury.com and
sign up for our newsletters.

Contents

Contents

Illustrations

Figures

Plates

Contributors

Andrea Avidad is a lecturer at the City University of New York in the Department of Film and Communication Arts. She is a PhD candidate in Cinema Studies at New York University. Her research interests focus on genealogies of film sound, affect theory and histories of listening in the nineteenth and twentieth centuries. She has published on non-linguistic sound in film and listening as ecological practice. Her most recent work is 'Deadly Barks: Acousmaticity and Post-Animality in Lucrecia Martel's *La ciénaga*' (2020).

Manuel 'Mandel' Cabrera Jr. is an assistant professor of philosophy at Underwood International College, Yonsei University in Seoul, South Korea. His major research interests are in philosophy of religion, philosophy and the arts, phenomenology and Spinoza's religious thought. He has published essays in metaphysics, philosophy of action, as well as in philosophy and film and literature, most recently 'Terrence Malick's Cosmic Cinema', forthcoming in *Life above the Clouds: Philosophy in the Films of Terrence Malick*, ed. Steven DeLay.

Aidan Delaney works in the Faculty of Arts and Creative Industries at Middlesex University, London. He teaches audio production and post-production to media students. He also manages the Media Department's technical facilities. Aidan's research interests are in sound studies, film studies, digital humanities, educational technology, remix studies and the philosophy of technology. He has published contributions on sound, film studies and remix culture. Aidan is an Avid-Certified Instructor in Pro Tools and has worked in the entertainment industry since the mid-1990s as a live sound engineer.

Jim Drobnick is a curator and professor of contemporary art and theory at OCAD University, Toronto. He has published on the visual arts, performance, the senses and post-media practices in anthologies such as *Olfactory Art and the Political in the Age of Resistance* (2021), *Food and Museums* (2017), *L'Art Olfactif Contemporain* (2015), *The Multisensory Museum* (2014), *Senses and the City* (2011) and *Art, History and the Senses* (2010). His books include the anthologies *Aural Cultures* (2004) and *The Smell Culture Reader* (2006). He is co-editor of the *Journal of Curatorial Studies* and co-founder of the curatorial collaborative DisplayCult (www.displaycult.com).

Charlotte Eubanks is a professor of comparative literature, Japanese, and Asian studies at the Pennsylvania State University. Research interests include literary Buddhism, the material culture of books, sound studies, and word/image relations from the ninth century to the present. They are the author of *Miracles of Book and Body: Buddhist Textual Culture and Medieval Japan* (2011) and *The Art of Persistence: Akamatsu Toshiko and the Visual Cultures of Trans-war Japan* (2020). Their articles have appeared in *Ars Orientalis*, the *Harvard Journal of Asiatic Studies*, *Book History*, *Postmedieval: A Journal of Medieval Cultural Studies*, *Verge: Studies in Global Asias* and other venues.

Andrija Filipović is an associate professor of philosophy and art and media theory at the Faculty of Media and Communications in Belgrade. They are the author of *Ars ahumana: Anthropocene Ontographies in the 21st Century Art and Culture* (2022) and *Conditio ahumana: Immanence and Ahuman in the Anthropocene* (2019), and monographs on Brian Massumi (2016) and Gilles Deleuze (2015). Their articles appeared in *Sexualities*, *The Comparatist*, *Contemporary Social Science*, *NORMA* and a number of edited volumes. Their research interests include environmental humanities, queer theory and contemporary philosophy. They are executive editor of *AM: Journal of Art and Media Studies*.

Jennifer Fisher is a curator and professor of contemporary art and curatorial studies at York University, Toronto. Her research focuses on exhibition practices, affect theory and the aesthetics of the non-visual senses. Her writings have been featured in publications such as *On Curating* (2021), *Linda Montano* (2017), *Caught in the Act, Vol. 2* (2016), *are you experienced?* (2015), *The Artist as Curator* (2015), *The Ashgate Research Companion to Paranormal Culture* (2013) and *The Senses in Performance* (2006). She is the editor of *Technologies of Intuition* (2006). She is co-editor of the *Journal of Curatorial Studies* and co-founder of the curatorial collaborative DisplayCult (www.displaycult.com).

Rob Garbutt is a writer and artist researching at the intersection of place, identity and belonging. Rob has published on settler colonialism, countercultures and contemporary Australian culture. His books include *The Locals* (2011) and *Inside Australian Culture* (2014). Rob's creative research engages with more-than-human environments. His most recent work, 'Fluencies' (2021), was exhibited in the exhibition *Speaking with the River*, Ballina, Australia. Rob is an adjunct associate professor in the Faculty of Business, Law and Arts at Southern Cross University.

Julius Greve is a postdoctoral research associate at the Institute for English and American Studies, University of Oldenburg, Germany. He is the author of *Shreds of Matter: Cormac McCarthy and the Concept of Nature* (2018), and the co-editor of *America and the Musical Unconscious* (2015), *Superpositions: Laruelle and the Humanities* (2017), 'Cormac McCarthy Between Worlds' (*EJAS*-special issue, 2017), *Spaces and Fictions of the Weird and the Fantastic: Ecologies, Geographies, Oddities* (2019) and *The American Weird: Concept and Medium* (2020). He is currently working on a manuscript that delineates the relation between modern poetics and ventriloquism.

Greg Hainge is a professor of French at the University of Queensland. A fellow of the Australian Academy of the Humanities, he has published widely in the areas of sound studies, film studies, literary studies, critical theory and more. He is the author of over fifty articles and chapters in collected volumes and three monographs, including *Noise Matters: Towards an Ontology of Noise* (2013) and *Philippe Grandrieux: Sonic Cinema* (2017). He is co-editor of the Bloomsbury Academic sound studies book series 'ex:centrics' and editor in chief of *Culture, Theory and Critique*.

Sandra Kazlauskaitė is a lecturer in sound and music theory at the University of Lincoln, UK. She was awarded a PhD at Goldsmiths, University of London, in 2019. Her research addresses feminist sonic practices, gendered soundscapes and the politics of space. As an artist and a curator, Sandra has worked across the disciplines of sound performance and installation art. Her publications include 'Soundscapes of the Post-Soviet World Today' in *Leonardo Music Journal* (2015), 'Doing Sonic Feminism' in *Atlas of Diagrammatic Imagination* (2019) and 'Women Sonic Thinkers. The Histories of Seeing, Touching and Embodying Sound' in *The Bloomsbury Handbook of Sound Art* (2020).

Runchao Liu specializes in critical media and cultural studies with a primary focus on Asian American popular music, musical orientalism and alternative activisms. Liu is an assistant professor in the Department of Media, Film, and Journalism Studies at the University of Denver and a visiting scholar at the Asian/Pacific/American Institute at NYU. Liu also serves as a co-editor of *Teaching Media Quarterly*. You may find her/their academic writings in *Cinéma & Cie*, *Critical Asian Studies*, *M/C: A Journal of Media and Culture* and forthcoming edited collections *Critical Race Media Literacy* and *The Cultural Politics of Femvertising*.

Sharon Jane Mee is an adjunct lecturer in film studies at the University of New South Wales, Australia. Her research applies poststructuralist and posthuman-feminist approaches to questions of the sensory and sensuous cinematic experience and the aesthetics and ethics of rhythm, movement and affect in experimental film and horror film. Her publications include *The Pulse in Cinema: The Aesthetics of Horror* (2020) and 'Thinking|Feeling Animality: Posthuman-Feminist Perspectives' in *The Routledge Companion to Gender and Affect* (forthcoming).

Norie Neumark is an honorary professorial fellow at the Victorian College of the Arts, Faculty of Fine Arts and Music, University of Melbourne, and emeritus professor, La Trobe University, Melbourne, and founding editor of *Unlikely: Journal for Creative Arts* (http://unlikely.net.au). As a sound/media artist and theorist, her research is mainly human-animal relations, environmental concerns and voice. Her collaborative art practice with Maria Miranda (www.out-of-sync.com) has been commissioned and exhibited nationally and internationally. Her writing on voice includes *Voicetracks: Attuning to Voice in Media and the Arts* (2017) and *Voice: Vocal Aesthetics in Digital Arts and Media*, (2010), lead editor and contributor.

Katherine Nolan is a lecturer in creative digital media at Technological University, Dublin. She is a practice-based researcher in contemporary art practice, especially media and performance. Her research investigates gender and embodiment in live and digital contexts. Recent solo exhibitions include *Fluid Flesh* (2021) and the *Mistress of the Mantle* (2017) at MART, Dublin. Recent papers include 'Fear of Missing Out: Performance Art through the Lens of Participatory Digital Culture' in the *International Journal of Performance Arts & Digital Media* (2021) and 'Life on Pause: Entanglements of the Maternal and the Mortal in a Global Pandemic' in the edited volume *Performance in a Pandemic* (2021).

Luke Robinson teaches Film Studies at the University of New South Wales, Australia. His PhD thesis on scenes of disappearing faces in 1940s Hollywood gothic cinema is currently being reworked as a monograph. Luke's research interests are in classical Hollywood film, film theory, politics of erasure, and theories of film sound. He is also a video artist working with Move in Pictures (https://www.move-in-pictures.com). Luke's second co-edited collection with Melanie Robson is called *One Shot Hitchcock: A Contemporary Approach to the Screen* and is due to be published by Oxford University Press in 2024.

Christian de Mouilpied Sancto is a PhD student in visual and cultural studies at the University of Rochester. He researches the histories of media technologies, architecture and urbanism, and resource extraction in the twentieth and twenty-first centuries. Previously, he studied music composition at Goldsmiths, University of London, and worked as a professional musician. His writing has been published in *Resonance, PAJ, Afterimage, Film-Philosophy* and *ASAP/Journal*.

Rachel Shearer is an artist and a lecturer based at *Te Wānanga Aronui o Tāmaki Makau Rau*, Auckland University of Technology in Aotearoa, New Zealand. Rachel's work and research investigate sound as a medium through a range of practices – art installation, composing/producing/performing experimental music, writing, as well as collaborating as a sound designer/composer for moving image and live performance events. Rachel is programme director for the Bachelor of Creative Technologies at Auckland University of Technology from where she also holds a PhD. Rachel affiliates to Rongowhakaata, Te Aitanga ā Māhaki *iwi* (extended kinship groups) and Pākehā (NZ European).

Acknowledgements

My inspiration for the energy that I put into this project was born out of four encounters with sound affects.

Encounter with sound affects #1

At the close of a long and arduous task resulting in a doctoral thesis titled 'A Sanguinary Economy and Aesthetics: The Pulse in Horror Cinema', I have the pleasure of receiving a response from a wonderful academic whom I greatly admire: 'I would like to see some consideration of sound as well.' The response is by no means surprising, for I had spent five chapters and a good 90,000 words outlining the rhythmic dispositions of the image and of the spectator with only brief mention of sound. This is to be my next evolution: pulse sounds.

Encounter with sound affects #2

In 2018, at a conference in Perth (Australia), I hear about Eve Klein's biomedical performance and installation work *Vocal Womb* (https://www .eveklein.com/vocal-womb) in a conversation with the artist over breakfast. In her installation work, Klein is wearing a corset and singing the repeated word 'contractions' in operatic tones, while the laryngoscope camera captures and projects the movement of her vocal cords on the screen. One is reminded of the contractions in giving birth and the ways that the sounds of the female body are all too often relegated to the order of silence. If Klein's *Vocal Womb* envisions the contractions of the larynx to produce vocal sounds, then for me, the practice of listening is a 'project of resistance' – a resistance to the ways that dialectical politics, including medical discourses, organize, regiment and silence female bodies.

Encounter with sound affects #3

Previous to my attendance at the conference in Perth, I encounter a laryngoscope in my curatorial work on the *This Is a Voice* exhibition at the Museum of Applied Arts and Sciences (MAAS). The laryngoscope was manufactured by the Down Bros. in London in 1890 and is representative of a history of laryngoscope production that began in 1829 when Benjamin Guy Babington invented what he called the 'glottoscope'. After this time, laryngoscopes were used by medical practitioners and vocal teachers alike to observe the movement of the vocal cords during the production of sound. One thing is for sure, the medical practitioners and vocal teachers, in their examination of the movement of the vocal cords, were not thinking outside of a system that organizes and regiments particular bodies for their discourse.

Encounter with sound affects #4

During the COVID-19 pandemic, I sit in a quiet room and hear a buzzing in my ears. I wonder about my body's resistance to silence and seek cures. I find that cures for tinnitus include this (somewhat painful) 'Tinnitus Cure Frequencies 100Hz – 23000Hz, Optimize Damaged Hearing' (https://www.youtube.com/watch?v=tVLsivMpQCQ). My 'project of resistance' to a system that silences, organizes and regiments flesh continues unabated.

For showing me how to think bravely, I thank Steven Shaviro, Eve Klein and Tilly Boleyn. Early work completed for 'The Clatter and Clang of Sound Affects' was published in the journal article 'Rhythm Beyond the Cinematic Medium/The Pixel Beyond the Movie Theatre'. I would like to thank *Cinéma & Cie* for permission to reprint material from this article. Jodi Brooks gave wonderful advice that was instrumental in the formative stages of the development of this collection. I would like to thank the authors in this volume whose intellectually impressive work inspired me to do better. I appreciate the time and effort that Luke Robinson put in to giving me feedback on my entry. I would like to thank Lisa Trahair for her tremendous energy, time and advice. Big thanks to Bernd Herzogenrath and Patricia Pisters for supporting the publication of this project. Many of our contributors had a difficult year; however, I am grateful to Katie Gallof at Bloomsbury for her wonderful editorial work, help and patience. I had many other supporters throughout this project: James P. Dahlstrom (love you to the moon and back), Gabrielle Lowe, Fiona Ransom, Lorna Mee and Michelle Mee. Lastly, I am indebted to Winston for his furry companionship and cheeky affection.

Sharon Jane Mee

When I was nineteen years old, my parents, artist Jude Robinson and poet Colin Campbell Robinson, took me to a Sydney cinema to see Bernardo Bertolucci's *Besieged* (1998). I was particularly transfixed by a scene where the composer (David Thewlis) asks the film's leading protagonist, Shandurai (Thandie Newton), why she does not return his love. At this moment, Shandurai cries out at the very same time that we hear the voice of Malian singer-songwriter, Salif Keita. What fascinated me the most about this scene was the emotive power of Shandurai's cry and how it was combined with Salif Keita's intensity of voice. Later, with Jodi Brooks's guidance, and having developed an understanding of Judith Butler and Michel Foucault's philosophies of power and Cathy Caruth on trauma, I had the opportunity to analyse *Besieged* through the lens of trauma, power, affect and sound theory. This was the genesis of my theoretical exploration of sound and affect. It was from that moment I started trying to think how a sound could be an affect and an affect could be a sound.

I pursued this theoretical interest in affect and sound in a research project on 'The Inaudible Sounds in Silent Cinema' with Sarah Cooper at Kings College London. I cannot thank Sarah enough for her exacting insights and feedback as I delved into this project, but also for her generosity towards me as a thinker while I was working with her. A move back to Sydney meant the pausing of the project; however, some of the research I undertook at this time into the relationship between sound and affect is presented here in this co-edited collection, including in sections of 'The Clatter and Clang of Sound Affects' and in my own entry on 'Inaudible Tremors'.

As a PhD candidate at the University of New South Wales, I have been very lucky to work closely once again with Jodi Brooks. Jodi is such a strong supporter of my intellectual pursuits and teaching practice. I am greatly privileged and wanted to express my deep gratitude to Jodi for her guidance, encouragement and feedback on my ideas and writing, which has made me not only a better film scholar but also a better person. I also want to thank George Kouvaros for all the support he has given me over the years at UNSW as an undergraduate, a PhD student and a casual academic. I must also express my deep thanks to Lisa Trahair for all the time she has given this project and for the advice she has shared along the way. Melanie Robson and Tara McLennan have been the most beautiful of friends and colleagues and their kindness and generosity (both intellectual and otherwise) have made many a day feel that much better. I have to thank my parents Jude and Colin for their words of wisdom and their love. A big thank you to George Damalas for his support and advice during the editing process. I would also like to thank Sharon Jane Mee for all the work and energy she has put into *Sound Affects*. Many warm thanks to all of the contributing authors with whom I have enjoyed much stimulating intellectual dialogue. I would like to thank Bernd Herzogenrath and Patricia Pisters for getting behind the co-edited collection and including it in their book series. And I am also very grateful for the editorial support of Katie Gallof at Bloomsbury. Finally, I would like to thank the Graduate School of University of New South Wales for paying me the Completion Scholarship and also the Australian Federal Government for providing me with the Australian Government Research Training Program Scholarship.

Luke Robinson

The clatter and clang of sound affects

Sharon Jane Mee and Luke Robinson

The title of this volume plays on the more generally recognized idea of *sound effects*. However, in *Sound Affects: A User's Guide*, we are not talking about the electric hum of Skaro – the home planet of the Daleks as shown on BBC TV – or the Wilhelm Scream that is used in countless films. Rather, as a collection we are providing a user's guide to *sound affects*. While this collection might not be *on* sound effects, it is very much a volume *of* sound affects. These sound affects range from those *felt*, 'heard' and repeated (silence, the oriental riff, shuffle), to the vocal (whispers, sing, the disembodied voice), to sounds at the threshold (tin/ny, thump, buzz), to sounds beyond the limits of audibility (inaudible tremors, distortion, sub-bass). As a volume of keywords, each of the entries in this collection conceptualize and theorize the relations between sound and affect. But, more importantly, in each of the entries, *a* sound affect (vroom), or the materiality of sound affects more generally (aa ee ii oo uuuuu, *squish, squelch, shlshlshlurpppp*), is described, exemplified and demonstrated through discussions of the material objects of the entry in question.

There are multiple sound affects, and, as a volume of keywords, *Sound Affects* is attentive to this multiplicity. In *The Forms of the Affects*, Eugenie Brinkema argues: 'There is no reason to assume that affects are identical aesthetically, politically, ethically, experientially, and formally; but only reading specific affects as having and being bound up with specific forms gives us the vocabulary for articulating those many differences.'[1] In this volume, we asked contributors to take up Brinkema's challenge and articulate each sound affect identified by closely engaging, describing and contextualizing a sound or audiovisual object. The objects chosen were not only to illustrate the concept in question but also to demonstrate how the object encourages us to rethink the relationships between sounds and affects. Before we introduce how *Sound Affects* is part of a landscape of film and media affect and sound theory, we, as we asked of the authors of this collection, will also articulate our conception of sound affects through an audiovisual example.

The 'form' of a sound affect: *Blacksmith Scene* (1893)

Blacksmith scenes have been central to the theorization of sound since approximately 500 BCE when the Greek philosopher Pythagoras began to develop his theory of musical intervals after visiting a blacksmith's forge (he is thought to have lived between *c.* 570 and *c.* 480 BCE).[2] It is rather fitting then to have the first object of this volume on sound and affect being an 1893 kinetoscope film known as *Blacksmith Scene*. *Blacksmith Scene* begins with a man lifting a piece of metal onto the anvil while two other men wait. Once the metal is placed on the anvil, two of the men, one on the left and the other on the right, begin hitting it with hammers. The first strike is from the man on the left, this is followed by a strike from the man on the right. The two men then continue working in a constant metronome-like rhythm. The man at the back of the anvil is armed with a smaller hand hammer. He begins by making three smaller hits at the side of the anvil, as if getting into the rhythmic beat, before he then joins the other two men at hitting the metal with a faster but less coordinated pace. The film contrasts the movement of the slow, thump-like hammer with the smaller hammer hitting out with quick beats, a contrast that is accented because the three men are hitting a still anvil. After the three men have hit the anvil for a while, they take a break and share a bottle of beer.

   Blacksmith Scene was the first narrative kinetoscope film shown to a selected public. This film was made on behalf of Thomas Edison, produced by the French-born William K. L. Dickson, and the cinematographer was the American director William Heise. The first public viewing of the film took place at the Brooklyn Institute New York on 9 May 1893.[3] The evening began with a presentation of stills from *Blacksmith Scene* projected by a magic lantern onto a large screen. Once the 400 or so projected still images were seen, the audience were asked to individually look into a kinetoscope and watch *Blacksmith Scene* move.[4] At the original screening of *Blacksmith Scene*, the president of the Brooklyn Institute's Department of Physics, George M. Hopkins, 'emphasized that the kinetoscope in its complete form was a machine for projecting images on the screen with the recorded synchronous sound'.[5] There was no sound accompanying the kinetoscope film, nonetheless, the audience was asked to infer the synchronous sound which, according to Hopkins, would be synchronized with it in the future. When K. L. Dickson and his sister Antonia Dickson later write about the film in their book on the kinetoscope, they describe the film as if a soundtrack had accompanied it. They say: 'set in their appropriate dramatic action the blacksmith is seen swinging his ponderous hammer exactly as in life, and the clang of the anvil keeps pace with his symmetrical movements.'[6]

In *Blacksmith Scene*, there is a forge similar to the one that is described by Pythagoras. The most-cited account of Pythagoras's visit to the blacksmiths is narrated by Iamblichus, a Neoplatonist philosopher from Cœle, now Syria. In his *On the Pythagorean Life*, a volume that was written in *c.* 245–*c.* 325 CE,[7] Iamblichus writes that Pythagoras 'walked passed [*sic*] a smithy, and heard the hammers beating out the iron on the anvil. They gave out a melody of sounds, harmonious except for one pair'.[8] Upon entering the forge and experimenting with the blacksmith's instruments, Pythagoras discovered that the different pitches of sound were created by the differing weight of the hammers.[9] According to Iamblichus, Pythagoras worked out that it was the weight of hammers that determine the quality of sound that is produced. Watching *Blacksmith Scene*, we are acutely aware of the force of movement that each differently weighted hammer has due to the alternate durations of time it takes for the hammers to go from a swing-up to a swing-down position. These different movements do not produce a sound, not even the 'clang', a sound effect that the Dickson's identify in their account of the film. These different movements do, however, reveal the potential for a sound to be made even if this sound is not actually made. It is this potential that we are suggesting led Hopkins to ask the audience of the inaugural presentation of the kinetoscope to imagine the sounds being produced as they watched the blacksmiths move. It is this potential that we argue constitutes a sound affect, a potential in the 'form' of vibrating sensation.

In *Blacksmith Scene*, the sound affect is the relation produced when the force – the pressure – of the hammer descends and meets a still anvil. It is the duration and intensity of this movement and impact that constitutes a sound affect, one that is not heard but nonetheless vibrates in, and across, the screen in the form of an inaudibly sonic pressure. Although the kinetoscope film is soundless (there is no synchronized soundtrack here), it conjures up sonority. Where Pythagoras believed that the movement of the heavenly bodies produced musical sounds heard by some,[10] such an idea has poetic, if not theoretical, resonance when watching *Blacksmith Scene*. While the film might not be accompanied by sound effects, it does nonetheless produce a sound affect in the form of pressure that is in itself a 'form' – one that is neither material nor immaterial, present nor absent,[11] but which nonetheless is intrinsic to how we as spectators come to enjoy this short film dramatizing a blacksmith scene.

The first audiovisual object of *Sound Affects* – *Blacksmith Scene* – is a film that has no soundtrack. It is, therefore, apropos that Michel Chion writes of all cinema: '*There is no soundtrack.*'[12] What Chion is referring to is the way in which a film cannot be studied independently of its sound.[13] For *Blacksmith Scene*, it is precisely the pressure – the force of intensity – of the hammers' beats that is 'heard'. What is in fact 'heard' is the sound affective force of the hammers, such that cinema is not image and soundtrack, but rather all the sensations of the scene by which we might describe sound affects. As Chion writes of cinema, this is 'an art in which sounds and images do not only seek to translate the strictly *audiovisual* world but all sorts of sensations as well'.[14] Indeed, for Chion, cinema is a 'sound art': an audiovisual 'simulacrum' that translates the sensory world.[15] The keywords of this volume help us identify the behaviour of the 'sorts of sensations' that are sound affects.[16]

While *sound effects* have the logic of a cause which produces an effect – with sound waves as the repercussion – *sound affects* can be described by their 'relations'. Notably, Bruce Baugh defines a 'body' in *The Deleuze Dictionary*:

> A body is not defined by either simple materiality, by its occupying space ('extension'), or by organic structure. It is defined by the relations of its parts (relations of relative motion and rest, speed and slowness), and by its actions and reactions with respect both to its environment or milieu and to its internal milieu.[17]

Similarly, defined not simply by their materiality, for *Blacksmith Scene*, the hammers that hit the anvil are characterized by their 'relative motion and rest, speed and slowness' and their 'actions and reactions' with respect to their environment.[18] Thinking beyond the 'relation' of the hammers to the anvil and to one another, one must also look to the 'relation' of the viewer with *Blacksmith Scene* to understand sound affects. As Felicity J. Colman writes: 'Affect is the change, or variation, that occurs when bodies collide, or come into contact.'[19] Thus, the affective force of the cinematic encounter is of 'affect as verbs becoming events . . . affects as perceivable forces, actions, and activities'.[20] To consider sound affects is to consider the moving and developing activity of sound as affect by which bodies – including viewer and screen – are engaged in 'dynamic transformations'.[21] Referring to the viewer's relationship to the cinematic apparatus for Dziga Vertov's kino-eye, Steven Shaviro writes: 'Cinema is at once a form of perception and a material perceived, a new way of encountering reality and a part of the reality thereby discovered for the first time.'[22] For the sound affective event, the inaudible sonic pressure that vibrates in and across the screen transforms the viewer as much as the metal is transformed by the weight, heft and force of the hammers of the blacksmith.

The sound affective event is also characterized by the viewer's cognitive encounter with it. According to Gilles Deleuze's cinema work on the movement-image, consciousness opens onto duration in the way that Henri Bergson in *Matter and Memory* aligns consciousness with duration (*durée*): 'it [consciousness] only existed in so far as it opened itself upon a whole, by coinciding with the opening up of a whole.'[23] In *Cinema 2: The Time-Image*, such an 'opening' in duration is the communicating vibrations in the *shock to thought* of movement.[24] As Deleuze writes: 'It is only when movement becomes automatic that the artistic essence of the image is realized: *producing a shock to thought, communicating vibrations to the cortex, touching the nervous and cerebral system directly.*'[25] What is apparent for particular forms of cinema that shock (Deleuze names the films of Dziga Vertov, Sergei Eisenstein and Abel Gance) is that they shock thought while imposing shock upon a mass-gathering public.[26] Deleuze writes: 'It is as if cinema were telling us: with me, with the movement-image, you can't escape the shock which arouses the thinker in you. A subjective and collective automaton for an automatic movement: the art of the "masses".'[27] For Deleuze, it is the 'automatic movement' of cinema that shocks:[28] a shock that was also experienced by the viewer when the kinetoscope allowed *Blacksmith Scene* to move. For *Blacksmith Scene*, the shock effect of the beat upon a surface – the force of the hammers hitting the anvil, as well as the force of intensities upon the consciousness of the viewer – is characterized by vibrations which communicate outward in relations. As Deleuze writes: 'From the image to the relation, and from the relation to the image: all the functions of thought are included in this circuit.'[29] The 'form' of the sound affect constitutes this type of circuit, where a *shock to thought* of the viewer is at the same time a shock *of* the image and, because it is a circuit, it is also a shock *to* the image.[30] We can see, in light of Deleuze's account of a *shock to thought*,[31] that when we are attentive to sound affects we begin to look and hear and think differently.

The theoretical work of our book thus takes up a Deleuzian notion of affect which intends a new formulation through sound. Our book also intends a new formulation via a *thinking* of sound affects. Characterized by sound affective and cognitive encounters, *Sound Affects* both 'thinks media' and reflects on our engagement with sonic and affective 'thinking media' to consider the ethical and political imperatives of such encounters. Just as via the 'automatic movement' of cinema and its status as mass-art, 'the movement-image was from the beginning linked to the organization of war, state propaganda, ordinary fascism, historically and essentially',[32] the medium of sound affects is infused ethically, politically, socially, historically and essentially. Thus, a thinking *with* and *reflecting on* sound affects is a thinking that has the force of sound affective and cognitive encounters by which the ethical and political powers of such sound affects ferment. As Colman writes about affect: 'affect describes the forces behind all forms of social production in the contemporary world, and these affective forces' ethical, ontological, cognitive, and physiological powers.'[33] Not all the audiovisual and/or sonic objects in this volume can be defined as mass-art, but they all adhere to a particularity which, when encountered by a viewer, may be indicative of a *shock to thought*.[34] In our book, sound affects describe the force of audiovisual and/or sonic objects – films, artworks, sound installations and music – by which we may consider the social, political, differential and ecological implications of sound via cultural artefacts; however, the force of such audiovisual and/or sonic objects may also be characterized by the 'powerlessness' of the 'unthinkable in thought'.[35]

The 'unthinkable in thought' takes on different qualities of expression in our volume. Where Deleuze describes 'the grey, the steam and the mist' in Akira Kurosawa's *Cobweb Castle* (also known as *Throne of Blood*, 1957) as the insensible, the imperceptible – as 'a thought, without body and without image'[36] – our book includes keywords which have the sonic and affective expression of a 'ventriloquism' heard in Kanye West's *My Beautiful Dark Twisted Fantasy* (Julius Greve's 'The Disembodied Voice') and the *squish, squelch, shlshlshlurpppp* of worms in compost ('*Squish, Squelch, Shlshlshlurpppp*' by Norie Neumark). Our book also includes keywords that illustrate the political implications of sonically affective works like *Whispering Campaign* (2017) by the African American artist Pope.L (Christian de Mouilpied Sancto's 'Whispers') and the sound affects of ecological impact ('Vroom' by Andrija Filipović). Inasmuch as such sound affects may not be graspable by the intellect, they are *felt* and 'heard' in the vibrations that ripple through the body, consciousness and the world. In this sense, such sound affects are graspable as 'carnal thought' in the way that Vivian Sobchack writes about the viewer's encounter with the 'sensations of audiovisual texture, color, and atmosphere' in Lucile Hadžihalilović's 2004 film *Innocence*: 'These are sensations rich with meaning but only really graspable in their richness not through intellect but through a form of "carnal thought" – "thought through my eyes", as Stephen Dedalus says in *Ulysses*'.[37] Sound affects, in the sense that they are *thought*, are vital to how we understand the force of audiovisual and/or sonic objects – as aesthetic, ethical and political ripples *felt* and 'heard' in the body, consciousness and the world.

Start, stop, repeat

This collection begins in silence – a soundless sound catapulted through space on the back of an inspired move by Katie Paterson to invite the moon to respond to a transmission from earth. Rewrite and revise: this *volume* begins in silence. *Volume*, in this sense, is the shape, capacity and breadth of the sounds and affects of bodies, ideas and conversations that this volume – as a collection or series of voices – accommodates. *Volume* has the 'form' of affect that Brinkema surveys in her book *The Forms of the Affects* – for this collection, the 'form' of *sound* affects – and of the sonic 'form' of the anechoic chamber that John Cage describes.[38] However, if the shape, capacity and breadth of the sounds and affects of bodies, ideas and conversations are indicative of a *volume*, we also find that bodies, as Jean-François Lyotard writes, are 'full of holes' – holes that are also involutions of the skin.[39] In such involutions of skin, the sonic and affective intensities of bodies – and indeed, ideas and conversations – pool and rest, vibrating along protrusions and causing ripples that vacillate and bloom outwards from the body, consciousness and the world. Thus, this *volume* begins in silence: a gentle stirring awake of voices, perhaps a casting of a stone that energizes those aesthetic, ethical and political ripples *felt* and 'heard' in the body, consciousness and the world, for the shape, capacity and breadth of sound affects comprised in this volume.

The first stone that is cast in this collection is Sandra Kazlauskaitė's entry, 'Silence', which tracks silence's relation to the historical and political discourses of powerlessness, disembodiment and inaction. Taking on this argument and aligning silence instead with embodiment and affect, Kazlauskaitė carefully traces the feminist discourses that allow for a way of understanding the affective tone and transformative possibilities of silence. Kazlauskaitė examines the 'form' of silence through sonic objects that include Katie Paterson's *Earth-Moon-Earth (4'33")* (2008) and Pauline Oliveros's set of *Sonic Meditations* (1971). The 'form' to which Kazlauskaitė refers is a 'sound/silence continuum' that is open to the potential affective realms of body and world.

Jim Drobnick and Jennifer Fisher's entry wonderfully examines the 'form' of shuffle through Christian Marclay's set of photographs on seventy-four playing cards, *Shuffle* (2007). In an exploration of a variety of sound affects – the action of shuffling the cards and the sounds and affects that result from the materiality and intensity of the handling, as well as the 'musical' psychogeographies of the cards – 'Shuffle' describes an embodied and affective form of play through musical negotiation and improvisation. The sound of the shuffling cards and the 'sound' of the musical notation for the player may be considered as both a bodily and cognitive encounter with the cards.

'The Oriental Riff' by Runchao Liu engages in political thinking about a musical trope of Eastern exoticism that has pervaded Western mediascapes for over a century. Liu examines the musical and repetitious 'form' of the oriental riff as aural witness to Asian racialization, while also engaging in a particular kind of 'disobedient listening'.[40] It is crucial to Liu's entry that Christine Bacareza Balance's concept of 'disobedient listening' provides both the theory and methodology for a practice that listens to the oriental riff in pop songs ranging from Carl Douglas's 'Kung Fu Fighting' (1974), Rush's 'A Passage to Bangkok' (1976), Siouxsie and the Banshees's 'Hong Kong Garden' (1978), David Bowie's 'China Girl' (1983) to Day Above Ground's 'Asian Girlz' (2013).[41]

Voices and vocals

In this section of *Sound Affects*, we turn to hear and feel the sound affects of the vocal tract. We begin with *aa ee ii oo uuuuu* of the *kōauau*, which, as Rachel Shearer tells us, is 'a flute from *te ao Māori* (the Māori world)'. Grounds and grounding are important concepts in Shearer's entry and in 'aa ee ii oo uuuuu', she lays the ground for a syllable – the vowel *aa ee ii oo uuuuu* – as sound affect. The ground is not purely physical or singular for Shearer. For her, there are multiple grounds, and means of grounding, for us to be attentive to. When the same idea is applied to this co-edited collection as a whole, each contribution can be considered as establishing other grounds (plural) for us to listen. For Shearer, the *aa ee ii oo uuuuu* of the *kōauau* resonates in and through time (the past, present and future), and also between the physical and spiritual, and the material and immaterial. With a focus on the resonance that occurs within these relationships, Shearer's argument in 'aa ee ii oo uuuuu' chimes perfectly with Charlotte Eubanks's entry that follows.

In Eubanks's contribution, we leave *te ao Māori* and find ourselves immersed in the Buddhist practices of Japan. Eubanks writes about the *kannō*, which translates as sympathetic response. In their entry they explain how '*kan* refers to the "stimulus" or "emotion" that a practitioner awakens in buddhas and *ō* to the "response" that this feeling generates in those beings'. A key object for their discussion is the *Gyosan taigaishū* manual, which Eubanks describes as 'a ritual handbook that has been used as the basic text for sutra chanting in the Shingon school of Japanese Buddhism since at least the twelfth century'. In this manual a musical system is transcribed, one that includes notes that cannot be produced by a human voice. However, as Eubanks argues, the 'human voice is directly keyed to the movement of the cosmos' and because 'the cosmos itself is made of "dust"' these vocally unproducible notes can possibly be seen when dust begins to dance. With Eubanks's contribution, as it is with many entries in *Sound Affects*, it is by paying attention to something that might otherwise go unnoticed – such as dancing dust – that a theory of a sound affect becomes a poetry of the unheard.

With Christian de Mouilpied Sancto's 'Whispers', we listen to that which may otherwise go unnoticed. Whispers can be defined as a type of vocal sound that is supposed to be heard . . . but only just. In his entry Sancto discusses Pope.L's *Whispering Campaign* – a sound installation made up of whispers that took place at documenta 14 held in Athens and Kassel during 2017. Core to Santo's argument is how '*Whispering Campaign* achieves multiple separations: of voice from source, of text from author, and of speech from affect.' From Sancto we, therefore, learn how sound affects are not always things, or forces, conjoining (sound with affect) but can also be forces where sound and affect are 'withdrawn' or 'withholding' from each other. Sancto's approach in his contribution opens new ways of thinking about affectless sound, or sound that is withholding affect. These concepts resonate with a potential (they are potential concepts), even if the potential of these concepts is punctuated by a force of its own negation.

The volume of sound is turned up with Katherine Nolan's 'Sing' or, actually, is it? since in 'Sing' we are invited to listen to – but also not listen to – 1980s and early 1990s power ballads that she performs in videos that have no synchronized sound. In Nolan's contribution she describes her practice-based methodology and establishes a ground for sing as a sound affect. She does so through a close discussion of two videos that are part of her *Silent Video Series: Fight the Rising Odds* (2015) and *Breathless* (2017). In her own words, 'the singing is energetic, exercised and shouted against the wind' in her 2015 video, while in the 2017 video: 'the voice is choked, held back and almost silent'. Crucial for her discussion of her own singing practice is how 'emotive sounds' come 'out of the body' while also resonating 'within the body as a form of self-affect'. Nolan then argues: 'As I perform the song, it performs me. Rather than simply an act of externalization, it is also an internalization of the song, an ingestion and a pulling inward.' As a ground, as sound affect, sing is simultaneously physical (it is of Nolan's body) and virtual (since the vocal movement is internal and external at the same time). The relationships between the physical and virtual are not only crucial for many of us who sing, the relationship between what is physical and what is virtual is also often a driving force for those who work with sound affects as a creative practice.

From the silent songs of Nolan's video practice, we come to the final entry of the 'Voices and Vocals' section of *Sound Affects*, which is Julius Greve's contribution on 'The Disembodied Voice'. He outlines his approach as follows: 'This entry . . . seeks to trace the maximally heterogeneous cultural and historical backdrop of the metaphoric of "ventriloquizing" in order to show the relevance of the affectively charged sound of the disembodied voice.' Greve then argues: 'The issue of ventriloquist speech also indexes debates within the humanities about what could arguably be called a vocal epistemology: an examination of the ways of knowing and expressing in and by one's own voice and those of others.' In his contribution, after a discussion of Kanye West's 2010 album *My Beautiful Dark Twisted Fantasy*, Greve establishes a conceptual base for a sound affect as a 'corollary' of the voiceless. We are at the end of the section on 'Voices and Vocals' but because Greve's 'The Disembodied Voice' gives voice to the voiceless, there is a conceptual bridge between his entry and the final section of *Sound Affects* – that of 'Beyond Audibility'.

Threshold sounds

For a *volume* of sound affects – where *volume* may refer to both a collection of written entries and the 'form' of sounds and affects – where are the boundaries? For the written text, are the boundaries defined by the covers of the book, the ideas and conversations that it contains, or the imagination of the reader? If the 'form' of sounds and affects may be defined by their shape, capacity and breadth – indeed, their characteristics and behaviours – as Brinkema writes about affect: 'the wild and many fecundities of specificity: difference, change, the particular, the contingent (*and*) the essential, the definite, the distinct, all dense details, and – again, to return to the spirit of Deleuze – the minor, inconsequential, secret, atomic',[42] are their boundaries bounded? Through the work of Deleuze and Félix Guattari, what we find is that bodily and cognitive forces, their intensities and capacity for movement, are defined not by boundaries but by 'thresholds'.[43] Thus, Deleuze and Guattari describe a body's bodily and cognitive encounter with other bodies by their 'capacities to affect and be affected'; a kind of 'acting on' and 'being affected' whereby bodies find their thresholds.[44]

Rob Garbutt's entry provides an astonishing way of thinking about thresholds through the sound affect 'Tin/ny'. In his entry, Garbutt describes the exterior surface of tin – as it is acted on by rain and wind – providing, what he calls, an 'extimacy' – or 'second skin' – for intimate affects.[45] It is crucial to Garbutt's conceptualization of 'extimacy' that tin brings forth affect (particularly anxiety) from the inside to the outside. Thus, through a political reading of a segment of the Opening Ceremony of the Sydney 2000 Olympic Games directed by Nigel Jamieson, 'Tin Symphony', and Warwick Thornton's film *Sweet Country* (2018) in which wind sonically activates the landscape and tin acts as the material for activation, Garbutt examines tin/ny's (non-Australians may need to read Garbutt's entry to understand the multivalency of the term) sound affective relation to post/colonial Australian identity and belonging.

'Vroom' by Andrija Filipović makes an admirable contribution through the sociopolitical conditions – and intensities – of a sound affective *vroom* in postsocialist Belgrade. Through the conceptually led methodology of ontosonography, Filipović engages in a considered 'graphing' of the intensities of the sounds of traffic and automobiles and their affective relation to Belgrade's streetscape. The 'maximal values' of the intensities of vroom are thresholds for sound and affect in the urban soundscape's *becoming*.

Andrea Avidad's 'Thump' examines the sonic and affective 'form' of significant thumps in Lucrecia Martel's film *La Mujer Sin* Cabeza/*The Headless Woman* (2008). Avidad contends that the 'form' of a thump is constituted through three essential terms: violence, mediation and dullness. Sound affective thresholds are found in 'the "thump-ness" of the thump'. In *La Mujer Sin Cabeza*, the violence of the thump – a sound that refuses to disclose the specificity of body – puts Vero (María Onetto) into a state of dullness and inanimation. Indeed, the notion of 'the headless woman' brings a literal kind of meaning to de-animated thought. Avidad's entry makes an outstanding contribution through the ethics of an affective 'thump', its relation to cognition and the class-inflected violence of a bourgeois consciousness.

In an examination of the buzzing fly in Paul Solet's film *Grace* (2009) and Darren Aronofsky's film *mother!* (2017), 'Buzz' by Sharon Jane Mee considers the sound affects that 'refuse [. . .] to go away' in cinema.[46] Mee contends that the buzzing fly is sonically and affectively undialectical; that is, as a sound affective spectre of the micro-movements and 'near inaudible' sonic intensities of decomposition,[47] the sonic and affective intensities of the fly are not recuperated in the meaning and silence of the gravestone or the filmstrip. The 'form' of buzz is thus understood through a dynamism that opens the viewer to relations with the corpse – and with the image – in an 'ontological fold'.[48] Such an 'ontological fold' opens the viewer to new sonic and affective worlds through which the body's capacity for sound affective intensities passes beyond thresholds.[49]

Beyond audibility

The entries in 'Beyond Audibility' conceptualize that which cannot be heard. The contributions also interrogate sounds and noises that can be heard but which somehow destabilize our relationship to what it is that we are hearing. Through his close analysis of three films directed by D. W. Griffith, Luke Robinson establishes inaudible tremors as a sound affect. For him a sound affect 'is a sound that is in the process of being actualized – a sound that is actualizing – rather than being a fully actualized sound'. With reference to two Griffith films, Robinson argues that inaudible tremors are an acoustic version of the tremors of the earth. If establishing a ground can help us be attentive to sound affects that might otherwise go unnoticed, such as the *aa ee ii oo uuuuu* of the *kōauau* or dancing dust, inaudible tremors are the manifestation of a rumble that vibrates through such groundings, demanding attention to what otherwise might be buried, or is soon to be buried, within this ground.

Like others in this co-edited collection Greg Hainge, in his contribution on 'Distortion', establishes a way of understanding sound and affect. And like others he does so in, and through, his discussion of a particular sound affect, in his case distortion. In his entry, Hainge argues: 'distortion will reveal to us how sound, when understood as an affect, can have no form, cannot be sound (in either the nominal or adjectival sense of this word) – and in doing so it will speak to us of much more than just sound.' For Hainge, distortion 'bridges the realms of the immaterial and material or, rather, conjoins them, demonstrating that no form is immutable, self-sufficient and self-identical across time'. Distortion for Hainge joins together what might often be considered separate, at the same time he argues that distortion instils difference where there might otherwise only be something (narcissistically) self-identical. Hainge also discusses and describes Low's 2018 album *Double Negative* and its reception. For him, 'the songs on the album, as many reviewers have commented, sound like nothing ever heard before, like an entirely alien form of sound that has come from somewhere else, and this goes not only for their sonic material but their structure too'. As a sound affect, distortion is something alien, something manifestly other, something that is so different that it de-structures or perhaps de-grounds the potential for close or easy listening. Or perhaps we could also say, distortion is a sound affect that dissonates the potential plenitude of sounds and music that are otherwise clearly audible.

Manuel 'Mandel' Cabrera Jr.'s 'Feedback' conceptualizes a sound affect where, as he says, 'sounding and listening, for act and affect, become fraught with Damoclean dangers.' Deeply inspired by the writings of the French composer Eliane Radigue, for him, like other contributions in this co-edited collection, feedback provides connections to the 'vibratory continuum' that is both audible and inaudible, a sound and an affect. Feedback is a mechanical form of distortion that signals the precarious nature of what Cabrera calls a zone – but it could also be a ground – from which we listen. In this regard, like Hainge's contribution on distortion, feedback destabilizes a listener's relationship to what can and cannot be potentially heard.

Following Cabrera's entry is Aidan Delaney's 'Sub-bass', a contribution that also describes the potential uncomfortable affects of listening and feeling, feeling and listening. Delaney describes such uncomfortable affects through a close discussion of Gaspar Noé's *Irréversible* (2002), a film where the sub-bass on the soundtrack potentially incites nausea in the film's viewer. Delaney argues: 'As noise, sub-bass vibration is unwanted and therefore troublesome, as music it presents fullness and enclosure to propel bodies to move in compliant resonance. In each case it is its intensity that gives sub-bass its physicality and affective force.' As a sound affect, sub-bass has rumblings that are so low that we might not be able to hear them, but it does not mean that sub-bass cannot be physically felt in ways that shape the ways we view and hear films such as *Irréversible*.

Concluding the section on 'Beyond Audibility' and also the co-edited collection as a whole is Norie Neumark who returns us to the earth to listen to the sounds of '*squish, squelch, shlshlshlurpppp*'. Neumark defines this sound affect keyword as 'the nearly inaudible sound of wormy compost . . . its more-than-human voice'. Like others in this co-edited collection, Neumark is invested in 'sounds that call out for a response'. Like the Buddhists in Eubanks's contribution, or the voiceless of Greve's, or the protagonists of Griffith's films in Robinson's entry, the worms of *squish, squelch, shlshlshlurpppp* call out for someone to respond. Ultimately, it is Neumark's own writing that is the response to the sounds of worms that without her discussion, and onomatopoeic language, would continue to remain unheard. It is through reading Neumark that we can hear the *squish, squelch, shlshlshlurpppp* of compost and, because we are being attentive to this sound affect, we are reminded that recycling is a type of return, a cyclicality, that helps give the earth, and the Earth, a potential future.

Notes

1   Eugenie Brinkema, *The Forms of the Affects* (Durham, NC and London: Duke University Press, 2014), xv.

2   Christoph Riedweg, *Pythagoras: His Life, Teaching, and Influence*, trans. Steven Rendall in collaboration with Christoph Riedweg and Andreas Schatzmann (2002; Ithaca: Cornell University Press; Bristol: University Presses Marketing, 2005), 27.

3   Charles Musser, *Before the Nickelodeon: Edwin S. Porter and the Edison Manufacturing Company* (Berkeley, Los Angeles and Oxford: University of California Press, 1991), 32; Charles Musser, 'A Cornucopia of Images: Comparison and Judgment across Theatre, Film, and the Visual Arts during the Late Nineteenth Century', in *Moving Pictures: American Art and Early Film,*

1880–1910, ed. Nancy Mowll Mathews with Charles Musser (New York: Hudson Hills Press, 2005), 5, 29. The film was later re-titled as either *The Village Blacksmith Shop* or *New Blacksmith Shop* and was also remade by the Lumière brothers as *Les Forgerons* (The Blacksmiths, 1895). Musser, 'A Cornucopia of Images', 5, 30. Musser cites Michelle Aubert and Jean-Claude Seguin, *La Production cinématographique des Frères Lumière* (Paris: Éditions Mémoires de Cinéma, 1996), 214.

4   Musser, *Before the Nickelodeon*, 35–6.
5   Musser, *Before the Nickelodeon*, 35–6.
6   W. K. L. Dickson and Antonia Dickson, *History of the Kinetograph, Kineto-scope, and Kineto-phonograph* (New York: Museum of Modern Art, 2001), 18.
7   Iamblichus's book was previously known as *Life of Pythagoras, or Pythagoric Life*, see Gilliam Clark's 'Introduction' to her translation of Iamblichus's text for the necessity of this change. Gillian Clark, 'Introduction', in *Iamblichus: On The Pythagorean Life*, trans. with notes and intro. Gillian Clark (Liverpool: Liverpool University Press, 1989), ix.
8   Iamblichus, *Iamblichus*, 51. Clark's translation is indebted to chapter 10 of Andrew Barker's then forthcoming edited volume, *Greek Musical Writings: Volume II, Harmonic and Acoustic Writings* (Cambridge: Cambridge University Press, 1990), 256–8. In this chapter, Barker presents fragments from the philosopher Nicomachus who probably lived in the second century. See Barker, *Greek Musical Writings*, 245.
9   Iamblichus, *Iamblichus*, 51.
10 Iamblichus, *Iamblichus*, 28. Also see Riedweg, *Pythagoras*, 29.
11 See Davina Quinlivan, *The Place of Breath in Cinema* (Edinburgh: Edinburgh University Press, 2012); Eugenie Brinkema, with composition by Evan Johnson, 'Critique of Silence', *Differences: A Journal of Feminist Cultural Studies* 22, nos 2–3 (2011): 211–34.
12 Michel Chion, *Film, A Sound Art*, trans. Claudia Gorbman (2003; New York: Columbia University Press, 2009), xi.
13 Chion, *Film, A Sound Art*, xi.
14 Chion, *Film, A Sound Art*, xi.
15 Chion, *Film, A Sound Art*, xi.
16 Chion, *Film, A Sound Art*, xi.
17 Bruce Baugh, 'Body', in *The Deleuze Dictionary*, ed. Adrian Parr, rev. edn (Edinburgh: Edinburgh University Press, 2010), 35.
18 Baugh, 'Body', 35.
19 Felicity J. Colman, 'Affect', in *The Deleuze Dictionary*, ed. Adrian Parr, rev. edn (Edinburgh: Edinburgh University Press, 2010), 11.
20 Colman, 'Affect', 13.
21 Steven Shaviro, *The Cinematic Body: Theory Out of Bounds 2* (Minneapolis and London: University of Minnesota Press, 1993), 41.
22 Shaviro, *The Cinematic Body*, 41.

23 Gilles Deleuze, *Cinema 1: The Movement-Image*, trans. Hugh Tomlinson and Barbara Habberjam (1983; Minneapolis: University of Minnesota Press, 1986), 10.
24 Gilles Deleuze, *Cinema 2: The Time-Image*, trans. Hugh Tomlinson and Robert Galeta (1985; Minneapolis: University of Minnesota Press, 1989), 156.
25 Deleuze, *Cinema 2*, 156.
26 Deleuze, *Cinema 2*, 157.
27 Deleuze, *Cinema 2*, 156–7.
28 Deleuze, *Cinema 2*, 157.
29 Deleuze, *Cinema 2*, 163–4.
30 Deleuze, *Cinema 2*, 156.
31 Deleuze, *Cinema 2*, 156.
32 Deleuze, *Cinema 2*, 165.
33 Colman, 'Affect', 13.
34 Deleuze, *Cinema 2*, 156.
35 Deleuze, *Cinema 2*, 166, 168. Deleuze invokes Antonin Artaud and a 'figure of nothingness' (Deleuze, *Cinema 2*, 168) when he writes: 'What cinema advances is not the power of thought but its "impower"' (Deleuze, *Cinema 2*, 166). However, Deleuze also invokes Jean-Louis Schefer when he writes, 'the cinematographic image, as soon as it takes on its aberration of movement, carries out a *suspension of the world* or affects the visible with a *disturbance*' (Deleuze, *Cinema 2*, 168). Such a '*disturbance*', 'far from making thought visible', has force (Deleuze, *Cinema 2*, 168).
36 Deleuze, *Cinema 2*, 169; quoting from Jean-Louis Schefer, *L'homme ordinaire du cinéma* (Paris: Cahiers du cinéma/Gallimard, 1980), 113–23, passim.
37 Vivian Sobchack, 'Waking Life: Vivian Sobchack on the Experience of Innocence', *Film Comment* 41, no. 6 (2005): 49.
38 John Cage, *Silence: Lectures and Writings* (Middletown, CT: Wesleyan University Press, 1973), 8.
39 Jean-François Lyotard, 'Several Silences', in *Driftworks*, ed. Roger McKeon and trans. Joseph Maier (1972; New York: Semiotext(e), 1984), 99.
40 Christine Bacareza Balance, *Tropical Renditions: Making Musical Scenes in Filipino America* (Durham, NC: Duke University Press, 2016), 9.
41 Balance, *Tropical Renditions*, 9.
42 Brinkema, *The Forms of the Affects*, xv.
43 Deleuze and Guattari write: 'All so-called initiatory journeys include these thresholds and doors where becoming itself becomes, and where one changes becoming depending on the "hour" of the world, the circles of hell, or the stages of a journey that sets scales, forms, and cries in variation' (Gilles Deleuze and Félix Guattari, *A Thousand Plateaus: Capitalism and Schizophrenia*, trans. Brian Massumi (1980; Minneapolis and London: University of Minnesota Press, 1987), 249).

44 Deleuze and Guattari, *A Thousand Plateaus*, 261.
45 Jacques Lacan, *The Ethics of Psychoanalysis*, trans. Dennis Porter (1986; London: Routledge, 2008), 139.
46 Steven Shaviro, 'Untimely Bodies: Towards a Comparative Film Theory of Human Figures, Temporalities and Visibilities', *SCMS* Conference, 2008, http://ftp.shaviro.com/Othertexts/SCMS08Response.pdf (accessed 27 November 2020).
47 Brinkema, 'Critique of Silence', 213.
48 Brinkema, *The Forms of the Affects*, 83.
49 Brinkema, *The Forms of the Affects*, 83.

Part I

Start, stop, repeat

Entry 1

Silence

Sandra Kazlauskaitė

Prologue

– Silence no. 1 –

In *Earth-Moon-Earth (4'33")* (2008), the contemporary artist Katie Paterson transmitted 4 minutes and 33 seconds of silence from Earth to the Moon with the hope that the celestial object would respond to the call. With reference to John Cage's emblematic composition *4'33"* (1952) – a silent piece conceived with no musical notes – Paterson's artwork repurposed the Cagean silence by opening it up to the cosmos, asking the Moon to listen and to respond. The silent signal travelled for approximately 500,000 miles before touching the planet's surface. The Moon sent a response to Earth 2.5 seconds later, and in the moment of their 'contact', silence became their shared language. Paterson's mixed media installation tapped into the possibilities of traversing space, discovering farther worlds through silence, while poetically revealing that silence might be far-reaching. Once released into the universe, in the case of *Earth-Moon-Earth (4'33")*, silence used its power to touch the Moon's surface, connecting us with the planet's luminous body.

– Silence no. 2 –

'Take a walk at night', the avant-garde composer and artist Pauline Oliveros instructs us. 'Walk so silently that the bottoms of your feet become ears.'[1] The composer's *Sonic Meditations*, produced in 1971, encourages us to turn to our bodies to listen to the full spectrum of sound, including silence. The composer believes that by allowing ourselves to connect with our internal bodily silences as well as silences of the outside world, we might discover new forms of intersubjective awareness and connectedness. Oliveros's compositions explore how silence may perform within one's body: the bottoms of our feet, the gaps and cracks in the mouth, the hidden workings of the bones, the stillness of our stomachs. By tuning our ears as well as our bodies to the sonically concealed and covered, we may discover that silence has its own timbre, shape and form. In order to locate the depths of what the composer refers to as 'the whole space/time continuum of sound/silences',[2] however, we ought to listen globally; we ought to commit to inclusive listening.

Introduction

Within the history of Western thought, silence has served as an ongoing contradiction, often finding itself in binary opposition to sound and provoking contrasting experiential intensities. On the one hand, to be in silence has meant to be in a space without sound, without voice, without presence – 'a vacuum, an airless, soundless void'.[3] As an 'apparent acoustic impossibility',[4] a mere 'absence of noise' and 'the cessation of speech',[5] silence has been repeatedly aligned with powerlessness rather than power due to its capacity to provoke isolation, exclusion and the loss of the self.[6] On the other hand, however, silence has been deemed 'good' for us. As a space for solitude and reflection, it presents a potential since it is believed to bring about 'a transformation of the self'.[7] If anything, Stuart Sim tells us, we need more silence as a way of combating the world's biggest problem – noise.[8] According to the author, where silence heals and rejuvenates, noise damages and destructs, this way reaffirming silence's attachment to predetermined binaries and separations.[9]

The paradoxical readings of silence only amplify the experiential complexity of the term. According to Mark Muldoon: 'silence is a slippery topic.'[10] While on a surface level silence may be read as an impossibility, 'these are only dispositions for the experience of silence'.[11] If anything, Ana María Ochoa Gautier argues: 'the tension between the apparent acoustic impossibility of silence and the intensely contrasting experiences it provokes lies at the heart of the types of presence and affect invoked by the term',[12] suggesting that silence inescapably contributes towards our affective positioning in the world. While overcoming silences, speaking up and finding a voice have been crucial to radical feminist practice, we have also learnt that silence is not a form of absence. It has a form – a history[13] – that can offer new imaginaries – new tools for speaking, sharing and being.[14] Hence, to be able to open silence towards its potential affective realms, it is important to consider the term beyond its paradoxes: beyond its binary framing against noise or beyond its alignment with lack and stasis. After all, by tuning in to the hidden and the covered tones of silence, we may discover, as Paterson's mixed media installation suggests, ways of connecting with new worlds, or we may also uncover, as Oliveros's practice encourages, collective presence.

This entry, thus, seeks to extend our understanding of silence and its transformative possibilities in affective and political terms. It considers how initiating, nurturing, listening to and embodying silences may contribute to the creation of what Deborah Kapchan calls 'the sound body'[15] and 'sound knowledge – a nondiscursive form of affective transmission resulting from the acts of listening'[16] through which binary discourses and dualisms, such as 'mind/body, nature/culture, man/woman, human/animal, spirit/material',[17] attached to capitalist and neoliberal structures, might be challenged. After all, if silence 'is not secondary to expressed thought but rather essential to embodied life',[18] it is important to consider how it contributes to the affective moulding of our day-to-day life. To explore the possibilities of the silent threshold, this entry reflects on the histories of aligning silence with experiential extremes and considers how silence, as a creative tool might help us to expand our understanding of the term. Specifically, it turns to Pauline Oliveros's silences, as practised in her *Sonic Meditations* works, and questions how opening our sound body to the full spectrum of silence may unearth new ways of being together and connecting to our lived environments in ways that disband binary systems and separations.

The different tones of silence

'Silence builds, never providing an autonomous offering. It is a site of multiple meanings.'[19] Historically, silence has served as a powerful metaphor to describe our place in the world. While capturing the experience of silence has not been fully possible, thinking about silence has continued to help us scrutinize the complex relationships between experience and power.[20] As a site of multiple meanings, silence has been conceptualized as both an impossibility ('[it is] not silence at all but sounds'[21]) and a material space – lived and experienced as real. According to Gautier, 'silence is lived as one of the most intense experiences across cultures.'[22] Silence can invoke pain or rage, lead to sensory deprivation, bringing us closer to near death. In its oppositional extreme, silence has the capacity to generate serenity and tranquillity. As a term, silence continues to present ambiguity, messiness and forcefulness that operates between the threshold of presence and absence as well as life and death. When considered within the spectrum of sonic experience, silence inevitably attaches itself to affect due to its capacity to *touch* our presence, *move* our bodies and *shape* our place in the world in its 'stubbornly unrecognizable'[23] form. Circulating 'somewhere on the scale that marks the entire range of sonic perceptions',[24] as a leaky vibrational flow, it diffuses in a 'feedback' loop between the inside and the outside: between bodies [in the broadest sense] and 'the fluctuations of feeling',[25] creating new experiential states and corporeal knowledge that go beyond what can be captured as feeling. Locating silence's affects, however, is not always an easy task – they can be refused, hidden or deployed upon bodies as 'the limit' in different ways.[26] And yet, silences continue to erupt in unexpected intensities and oscillations, making our bodies feel and react.

In the context of politics, for example, silence has been addressed in relation to power, examined through domination, non-participation and resistance, often attaching itself to negative affects. Within Western thinking, it has become 'common sense' to theorize silence as a disembodying state that manifests in powerlessness and voicelessness.[27] Indeed, silence can be silenc*ing* – it oppresses, it refuses expression, it immobilizes. It can be 'understood as the opposite of "having a voice," where voice is rendered as a sign of identity and presence of the subject'.[28] Without a voice, your subjectivity becomes negated, and your body, as a result, may become erased. The work of second-wave feminism, for example, exposed the violence that silence carries. In Adrienne Rich's words: 'In a world where language and naming are power, silence is oppression, is violence'.[29] We have witnessed radical feminists, including Audre Lorde, calling for silence to be transformed into language and action:

> and when we speak we are afraid
> our words will not be heard
> nor welcomed
> but when we are silent we are still afraid
>
> so it is better to speak remembering
> we were never meant to survive.[30]

bell hooks confronted silence by asking: 'how can we organise to challenge and change a system that cannot be named?'[31] The work of feminism has demonstrated that silence, located under patriarchal conditions, must be overcome in order to possess 'both the opportunity to speak and the respect to be heard'.[32] Contemporary debates in sound and cultural studies further reveal how the sinister dimension of silence continues to serve as a successful 'auditory mechanism of control'[33] within oppressive patriarchal structures. From the torturous silences of solitary confinements and extreme isolation to the eerily silences of abandoned histories of communities and peoples, as well as economic neglect of populations and environments,[34] the affective politics of denial, negation and torture have and continue to be entrenched in silence.

Within the histories of Western sonic thinking and practice, silence has been claimed as both an impossibility[35] and a human right.[36] The Futurists, for example, actively dismissed the silence of ancient life. The Futurist Manifesto claimed that 'with the invention of the machine, Noise was born',[37] saving the world from the muted tones of the past. For Cage, the experience of silence was unfeasible. After visiting the anechoic chamber at Harvard University in 1951, where he heard his body speak, Cage claimed that 'there is no such thing as an empty space or an empty time. There is always something to see, something to hear. In fact, try as we may to make a silence, we cannot.'[38] If anything, according to Cage, silence is all the noise that surrounds us. R. Murray Schafer's polemical study on soundscapes, on the other hand, aligned silence with positive affectivity, situating silence against noise. Schafer believed that silence found in nature is good for us, while urbanized noise is destructive to our health and bodies. As Marie Thompson comments: 'In Schafer's account, silence is equated with tranquillity; tranquillity is equated with the natural; and the natural is equated with the good',[39] while noise, as an element of culture in Schafer's view, is detrimental to our well-being. The 'natural' silence, Schafer promises, will clean our ears, and with that our bodies, which is necessary to our survival. For Schafer, Thompson notes, silence 'has the power to rejuvenate the body, mind and soul of the listening subject',[40] as well as free our bodies from noisy inhibitions and its affects. If untreated and left to noise, however, as Schafer warns, our perspective may be lost, and our surroundings may become disordered. In this sense, Schafer idealizes silence, suggesting that 'underneath the clamour of the perceptible soundscape lies an absolute, unbroken and ideal silence'.[41] Thompson critiques Schafer's silence/noise binary split by arguing that his romanticized longing for a 'more natural' past presents a problematic conservative politics of silence,[42] underlined by, as Annie Goh further contends, a 'white, masculinist patriotism'.[43] Thompson actively breaks out from the silence/noise dualism by arguing that silence is not inherently 'good' or 'bad' – the negative or positive understanding of silence is contingent upon its context and relations, as sound/silence continuum affects and effects unfold in situ.[44]

Here, the following question arises: if we move beyond the ingrained binaristic attachments associated with silence, where might we end up? Recent feminist accounts have invited us to reconsider silence as 'a space of fluidity, non-linearity' that communicates 'deeply at the edges of sound',[45] this way directing us towards the transformative possibilities of silence.

It is in this precise conceptual space that we might begin to uncover the affective tone of silence, one that circulates between environment and 'the sound body' – 'a resonant body that is porous, that transforms according to the vibrations of its environments, and correspondingly transforms that environment'.[46] As argued by Gust Yep and Susan Shimanoff, silence, as an expressive and embodied activity, is able to communicate 'a wide range of experiences, social relations, and cultural conditions'.[47] This entry turns to Oliveros's *Sonic Meditations* practice and considers how opening the affective field of circulation of feedback where silences pulsate and communicate might help us to unearth new states of being, awareness and embodied-political collectivity.

The transformative possibilities of silence in Pauline Oliveros's *Sonic Meditations* (1971)

Silence has played an important role within the arts, including music, film, literature and sound art.[48] Serving as an 'acoustic gap',[49] a blank canvas,[50] concealed noise, a quietus or heavenly apotheosis, artists have repeatedly turned to the metaphor of silence to express divinity, confront death and explore spaces between words, this way alluding to the literal as well as the metaphysical intensities of the term. According to Sim, silence within the arts has been used as a structural device and a thematic tool,[51] through which the so-called absences of sound could be conceptually questioned and explored. Whether it is the professed heavenly silence narrated in the *Book of Revelation*, the eerie silence in Ad Reinhardt's *Abstract Painting* (1963), the silent death in Ingmar Bergman's film *The Seventh Seal* (1957) or silence as *all* sound in John Cage's *4'33"* composition (1952), silence, as mediated through the arts, has continued to provoke uncanny experiential moments and 'touch' the listener in different ways. As Paterson reveals in her *Earth-Moon-Earth (4'33")* installation piece, it may even connect us to new planets.

In the context of the twentieth-century sonic experimentalism, for example, the compositions of Oliveros reveal how 'the binary logic of apparent opposites [equated with silence]' can be broken down 'by dissolving one into the other (presence as absence, emptiness as plenitude, quietness and expressivity, silence as intensity of life)'.[52] Oliveros's works actively steer away from the binaristic framing of silence. Instead, her practice opens silence towards the affective social milieu. In her writings and creative works, Oliveros considers silence as a lived presence that lies within the spectrum of sound and its affective possibilities. While for Cage, for example, experiencing silence was not possible, for Oliveros, silence is material; it is integral to the infinite spectrum of the sonic and its experiences. After all, as Kapchan writes: 'even when inaudible, [sound] is indelibly material. As a vibration, it permeates everything.'[53] Oliveros's practice implies that when we hear sound, we also hear silence. And when sound and silence circulate, they permeate everything together: 'There is no sound without silence before and after. Sound/Silence is a symbiotic relationship. Sound and silence are relative to one another.'[54]

Sound/silence, the composer's works convey, is a continuum that *lives* inside and outside our bodies as an unceasing event. As a vibration, it travels and moves, this way connecting and transforming bodies and environments over time. Opening our ears, our bodies as well as our environments to this precise continuum may help us to unveil the potential affective shape and form that silence carries within the social and political realm. As a material force and a shared practice, experienced within our bodies and the world, silence may manifest itself as a space for reflection, where collective power and agency can be created. Simultaneously, it may also open itself in the shape of resistance and possibility, where tools for connecting, listening and sharing beyond language can be unearthed. In order to locate the timbre, shape and form of silence, Oliveros teaches us, we must listen to the bodily gaps and environmental in-betweens, this way uncovering the affective depths that lie within the sound/silence continuum shared by communities.

For Oliveros, sound/silence is not an isolated construction but a shared space and a connecting practice. It is an act of collaboration. As a continuum, it circulates in the world always as both, subjective *and* intersubjective – always within the body, but also always belonging to the universe. While resonating between the inside and the outside, between bodies and their environments, sound/silence creates and conducts resonating affects, which, in circulation and feedback, pave the way to new potential collective imaginaries built on openness, fluidity and non-linearity, rather than predetermined binaries and oppositions.

These affects, as *Sonic Meditations* reveal, are open-ended and subject to ongoing discoveries. For example, Oliveros's *Sonic Meditations* actively invite affective transmissions between lived bodies and the 'sonosphere'.[55] The compositions encourage the participants to tune in to the seismic oscillations between our physical lived bodies – the workings of chemicals, nerves and hormones, and the rhythms of their environments.[56] Conceptually, the project utilizes silence as a communication tool, through which embodied social intersubjectivities between the listener and their community are encouraged to emerge as a shared event. Initiated as open-ended auditory experiments led by improvisation and bodywork, *Sonic Meditations* invites the participants to engage in steady awareness and attention exercises, to create sound/silence individually and communally. For example, in *Sonic Meditations XIII*, we are invited to 'listen to the environment as a drone. Establish contact mentally with all of the continuous external sounds and include all of your continuous internal sounds, such as blood pressure, heartbeat and nervous system.'[57] This improvisation exercise asks us to focus on our awareness, actively tune in to our bodies and listen to the inner and outer spaces between sounds as a way of accessing new collective meanings.

As Julia R. Johnson confirms, the body full of silence is a body full of possibilities. To discover its inner harvest and its social power, it is important to open up and listen to the bottoms of our feet, the hidden workings of the bones, the stillness of our stomachs as well as the external silences that surround it:

> Silence is accessible through the body. Moving into silence is a performance – a choice – to access new meanings and new realities. In our bodies, we can attend to the fuzzy edges of sound. . . . Intentionally attending to our bodies, we can feel and disrupt the linear timetables that often define daily life. Therein exists the potential to feel/know/hear/ perceive ourselves and each other more deeply and carefully.[58]

Sonic Meditations compositions could be considered as active steps towards opening up the possibilities of silence to access its affective potential – communal and intersubjective – through the acts of listening to our bodies. According to Stephen Miles, the works open us to 'the intersubjective world of social relationships'.[59] By actively engaging in inner and outer sound/ silences in Oliveros's *Sonic Meditations*, the participants bypass passive forms of aesthetic absorption, this way entering the sphere of social connectedness built on inclusive shared listening and shared affects. After all, if listened to inclusively, Oliveros proposes, sound/silence continuums can expand our affective, social and political consciousness.[60] *Sonic Meditations* teach us that opening silence to the infinite spectrum of the sonic, as experienced intersubjectively, contributes towards the cultivation of communities built on understanding, openness and compassion.

Conclusion

The analysis of Oliveros's *Sonic Meditations* reveals how silence is integral to our critical understanding of sound and its affective possibilities. By situating silence within the spectrum of the sonic, Oliveros invites the listeners to consider silence beyond its equation with voicelessness and disembodiment. By tuning in to and opening to the sound/silence continuum, we discover that silence, in its fluid shape and form, may offer resistance, rather than non-participation. In Oliveros's works, we are encouraged to actively open our sound bodies to the edges of sound where affects may be covered or hidden, where we might discover that a body full of silence is not an isolated disembodied body but an active site of agency, fuelled with intense, healing and connecting affects. After all, as argued by Malhotra and Rowe, the practice of silence collectively actives us. It 'allows us the space to breathe. It allows us the freedom of not having to exist constantly in reaction to what is said. Standing in silence allows for that breadth, for that reflection that can create a space of great healing'.[61] This practice is work. The fruits of this work, however, are transformative, and as Oliveros proposes, practice only enhances openness.[62] To unleash the full potential of silence in its affective and political terms, we are invited to practice, and to practice, openly. While reflecting on *Sonic Meditations*, Oliveros suggests that we can begin to create spaces for collective breath, healing, agency and resistance, once we develop a common bond with other bodies and environments through listening to sound/silences collectively. This becomes possible 'through a shared experience, when one's inner experience is made manifest and accepted by others, when one is aware of and in tune with one's surroundings, when one's memories, or values, are integrated with the present and understood by others'.[63] Sharing bodily and environmental silences can help us to disrupt linear systems, communicate deeply and access new imaginaries within the edges of sound. *Sonic Meditations*, as a creative practice of silence, reminds us that our bodies, in ongoing cyclical interconnection and coextension with our lived environments, are full of reciprocal affects with transformative sociopolitical power that can be communicated through and with silence. By opening spaces between worlds collectively, we can unloosen the knotted dualisms and create spaces where we can breathe, resist, heal and build affective solidarity – a conceptual basis for political change and transformation.[64]

Notes

1 Pauline Oliveros, *Sonic Meditations: March–November 1971* (Urbana, IL: Smith Publications, 1974).

2 Pauline Oliveros, *Deep Listening: A Composer's Sound Practice* (New York: Universe, 2005), xxiv.

3 Sean Street, *Sound Inside the Silence: Travels in the Sonic Imagination* (Cham: Palgrave Macmillan, 2020), 3.

4 Ana María Ochoa Gautier, 'Silence', in *Keywords in Sound*, ed. David Novak and Matt Sakakeeny (Durham, NC and London: Duke University Press, 2015), 184.
5 Mark S. Muldoon, 'Silence Revisited: Taking the Sight out of Auditory Qualities', *The Review of Metaphysics* 50, no. 2 (1996): 275.
6 Gautier, 'Silence', 183.
7 Gautier, 'Silence', 183.
8 For more, see Stuart Sim, *Manifesto for Silence: Confronting the Politics and Culture of Noise* (Edinburgh: Edinburgh University Press, 2007).
9 Sim, *Manifesto for Silence*, 4.
10 Muldoon, 'Silence Revisited', 275.
11 Muldoon, 'Silence Revisited', 275.
12 Gautier, 'Silence', 184.
13 Adrienne Rich, *On Lies, Secrets, and Silence* (New York: W. W. Norton & Company, 1979).
14 Sheena Malhotra and Aimee Carrillo Rowe, *Silence, Feminism, Power: Reflections at the Edges of Sound* (New York: Palgrave Macmillan, 2013), 1.
15 The entry follows Kapchan's reading of the body. The author identifies the body as a multitude of the physical and lived – an assemblage of organs, tissues as well as 'perfusion of hormones, chemicals, synapses, nerves; a plethora of molecules shedding and spreading beyond the skin, a substance that responds to the rhythms of its environment. The body: a site of recognition, an evanescent materiality, a pliant ambiguity' that transforms in time, as it is never complete. While it moves and feels, it also *sounds*. For more, see Deborah Kapchan, 'Body', in *Keywords in Sound*, ed. David Novak and Matt Sakakeeny (Durham, NC and London: Duke University Press, 2015), 34.
16 Kapchan, 'Body', 34.
17 Kapchan, 'Body', 34.
18 Malhotra and Rowe, *Silence, Feminism, Power*, 5.
19 Rachel Levitt, 'Silence Speaks Volumes: Counter-Hegemonic Silences, Deafness, and Alliance Work', in *Silence, Feminism, Power: Reflections at the Edges of Sound*, ed. Aimee Carrillo Rowe and Sheena Malhotra (New York: Palgrave Macmillan, 2013), 68.
20 For more, see Kris Acheson, 'Fences, Weapons, Gifts: Silences in the Context of Addiction', in *Silence, Feminism, Power: Reflections at the Edges of Sound*, ed. Aimee Carrillo Rowe and Sheena Malhotra (New York: Palgrave Macmillan, 2013), 188–99, and Kris Acheson, 'Silence as Gesture: Rethinking the Nature of Communicative Silences', *Communication Theory* 18, no. 4 (October 29, 2008): 535–55.
21 John Cage, *Silence: Lectures and Writings* (Middletown, CT: Wesleyan University Press, 1973), 22.
22 Gautier, 'Silence', 183.
23 Marie Thompson and Ian D. Biddle, *Sound, Music, Affect: Theorizing Sonic Experience* (New York: Bloomsbury Academic, 2013), 6.

24 Muldoon, 'Silence Revisited', 277.
25 Thompson and Biddle, *Sound, Music, Affect*, 6.
26 Gautier, 'Silence', 184.
27 Malhotra and Rowe, *Silence, Feminism, Power*, 1.
28 Gautier, 'Silence', 183.
29 Rich, *On Lies, Secrets, and Silence*, 204.
30 Audre Lorde, *The Collected Poems of Audre Lorde* (New York and London: W. W. Norton & Company, 2017).
31 bell hooks, *The Will to Change: Men, Masculinity, and Love* (New York: Washington Square Press, 2004), 25.
32 Lana Rakow and Laura A. Wackwitz, *Feminist Communication Theory: Selections in Context* (Thousand Oaks, CA; London; New Delhi: Sage Publications, 2004), 9.
33 Marie Thompson, *Beyond Unwanted Sound: Noise, Affect and Aesthetic Moralism* (New York; London; Oxford; New Delhi; Sydney: Bloomsbury Publishing, 2017), 102.
34 For more, see Elizabeth A Povinelli, *Economies of Abandonment: Social Belonging and Endurance in Late Liberalism* (Durham, NC: Duke University Press, 2011).
35 Cage, *Silence*.
36 R. Murray Schafer, *The Soundscape: The Tuning of the World* (Rochester, VT: Destiny Books, 1993).
37 Luigi Russolo, *The Art of Noise (Futurist Manifesto, 1913)*, trans. Robert Filliou (New York: Something Else Press, 1967), 23.
38 Cage, *Silence*, 7.
39 Thompson, *Beyond Unwanted Sound*, 92.
40 Thompson, *Beyond Unwanted Sound*, 95.
41 Thompson, *Beyond Unwanted Sound*, 97.
42 Thompson, *Beyond Unwanted Sound*, 88.
43 Annie Goh, 'Sounding Situated Knowledges: Echo in Archaeoacoustics', *Parallax* 23, no. 3 (July 3, 2017): 285.
44 Thompson, *Beyond Unwanted Sound*, 103.
45 Malhotra and Carrillo, *Silence, Feminism, Power*, 2.
46 Kapchan, 'Body', 38.
47 Gust A. Yep and Susan B. Shimanoff, 'The US Day of Silence: Sexualities, Silences, and the Will to Unsay in the Age of Empire', in *Silence, Feminism, Power: Reflections at the Edges of Sound*, ed. Aimee Carrillo Rowe and Sheena Malhotra (New York: Palgrave Macmillan, 2013), 141.
48 Sim, *Manifesto for Silence*.
49 Gautier, 'Silence', 184.
50 Sim comments how Kasimir Malevich's *White Square on a White Background* (1918) and Alexander Rodchenko's *Pure Red, Pure Yellow and Pure Blue* (1921), for example, could be considered as the painterly equivalent of silence, or 'extensions of silence'. For more, see Sim, *Manifesto for Silence*, 9–10.

51 Sim, *Manifesto for Silence*, 111.
52 Gautier, 'Silence', 184.
53 Kapchan, 'Body', 42.
54 Oliveros, *Deep Listening*, 14.
55 Sonosphere, according to Oliveros, is: 'all sounds that can be perceived by humans, animals, plants, trees and machines'. For more, see Pauline Oliveros, 'Auralizing in the Sonosphere: A Vocabulary for Inner Sound and Sounding', *Journal of Visual Culture* 10, no. 2 (August 2011): 163.
56 Kapchan, 'Body', 34.
57 Oliveros, *Sonic Meditations*.
58 Julia R. Johnson, 'Qwe're Performances of Silence: Many Ways to Live "Out Loud"', in *Silence, Feminism, Power: Reflections at the Edges of Sound*, ed. Aimee Carrillo Rowe and Sheena Malhotra (New York: Palgrave Macmillan, 2013), 63–4.
59 Stephen Miles, 'Objectivity and Intersubjectivity in Pauline Oliveros's "Sonic Meditations"', *Perspectives of New Music* 46, no. 1 (2008): 6.
60 Oliveros, *Deep Listening*, 16.
61 Malhotra and Row, *Silence, Feminism, Power*, 2.
62 Oliveros, *Deep Listening*, xxv.
63 Oliveros, *Sonic Meditations*.
64 Clare Hemmings, 'Affective Solidarity: Feminist Reflexivity and Political Transformation', *Feminist Theory* 13, no. 2 (August 2012): 147–61.

Entry 2

Shuffle

Jim Drobnick and Jennifer Fisher

Standard musical notation offers instructions for how performers should play and shape the affect of a composition. Symbols, marks and a host of terms direct aspects such as tone, pitch, breath, tempo, dynamics and articulation that generate the distinctive feeling sought by the composer. What happens, though, when performers are given an atypical score, with no conventional understanding about how the information is to be musically interpreted? *Shuffle* (2007), by Christian Marclay, an internationally renowned artist working in audio, video, print, performance, sculpture and installation, offers a compelling interrogation of music, improvisation and cross-modal sensory affects. Instead of a linear score, the artist provides performers with a series of photographs documenting musical notation found in everyday life and mounted onto seventy-four 16.8 × 12.1 centimetre cards (see Plate 1). The images feature recognizable clefs, notes and staves, encountered by the artist on umbrellas, beer coasters, street signs, interior decor, clothing, automobiles and other quotidian objects. These visual fragments of musical notation, discovered in Marclay's travels and captured in street-style photography, yield a near infinity of sequences that can be deliberately arranged or chosen by chance.

The title of the deck, *Shuffle*, employs a word with multivalent significance. It operates as both a noun and a verb, and a sound and an action, which has implications for both the music produced through the cards and their affective affordances. Dictionaries offer two main definitions of 'shuffle': a type of walk or dance step, and a rearranging or intermixing of a deck of cards. Both carry sonic implications, as the dragging of the feet produces a scraping or scuffing, and the jumbling of cards creates a papery rustling that can be soft or sharp depending on the intensity of the handling. There exists a third definition of shuffle, labelled 'archaic': behaving in a vague, evasive, equivocating manner.[1] To give someone the 'shuffle' foregrounds an action of getting out of an obligation, which, in this case, encompasses the artist's shifting of norms and conventions of reading musical notation in linear form. For collaborating musicians, *Shuffle* welcomes improvisation that requires navigating and sonically interpreting the 'music' appearing on the cards as well as the uninscribed gaps in between. This entry explores how Marclay's *Shuffle* serves as a compelling pivot to bring forward ideas about the contingencies between art, music and affect. Not only is the artist's work aptly named for the mixed modalities it utilizes, the title is significant in how the deck articulates and rearticulates the distinctions between the visual and the sonic, art and music, composer and performer, improvisation and notation.[2]

Shuffle continues a nearly 500-year history of music card decks. These decks carry suits and numbers, as traditional cards do, yet are accompanied by various lengths of musical notation: from single notes to several bars to entire songs. Peter Flötner's *spielkarten* (1535–40) features ribald imagery and humorous folk songs about fools, peasants, aristocrats and swine. Players can sing bawdy and moralizing tunes to match the themes pictured on the cards, such as two women fighting, expressing the discord that gambling tends to breed.[3] Decks can also connect to specific musical performances and well-known composers. The cards published as souvenirs of John Gay's *The Beggars' Opera* (1728) divided the music into fifty-two sections, each outlining a solo by one of the characters. Owners of the deck could thus recreate key scenes in the satirical opera wherever and whenever they chose.[4] A third genre of music cards is dedicated to instruction. Produced by educators and conservatories, these decks carry snippets of music to teach musical definitions, theory and techniques, as well as to offer playful opportunities for students to practice their *métier*.[5] Marclay's deck shares aspects of these genres of music cards in that it coincides with popular sensibility (music occurring in everyday life) and aligns with democratic availability (anyone can perform the cards). But there are striking differences. *Shuffle* notably avoids literal interpretations, thwarts a reductive connection to any single piece of music, and eschews didactic purposes. While such conceptual openness characterizes a large proportion of contemporary art, in Marclay's work, affective variability is used as a prompt to spark improvisation.

Within the history of art, precursors to *Shuffle*'s improvisation-inducing cards include those used by composers and musicians following the precedent of John Cage's experiments with the tarot and I Ching in the 1950s.[6] These aleatory procedures aimed to decentre the habits of creation, liberate the performer from the authority of the composer, and serve as a tool to generate novel sonic experiences. Proponents of new jazz, such as Peter Brötzmann, and members of Fluxus, such as Gianni-Emilio Simonetti, employed self-published decks as a randomizing strategy to cultivate indeterminacy and unexpectedness.[7] Combining text, drawing, collage, diagrams, images and symbols, sets of cards by Brötzmann (*Signs* and *Images*, both 2002) and

Simonetti (*Partita – Between Noise and Silence,* 2013) offer idiosyncratic visual cues to prompt unique and varied improvisational performances.

Marclay's cards differ by focusing solely on photography. The photographs, notably, were taken long before Marclay decided upon the deck format.[8] Even though *Shuffle* features musical symbols, these are all embedded in the milieu captured in the snapshot. While the artist has framed and selected the images that comprise the deck, the 'music' sourced for each card was not originally designed to produce a performance. Musicians using the cards are invited to enact a 'shuffle' free from any obligation to a composer's particular objective. Most of the musical phrases on the cards were found serendipitously and designed as décor or embellishments. Released from second-guessing what Marclay might have intended by the notations, performers interpret the photographs spontaneously and intuitively, rather than trying to 'read' them for specific directions.

Playing cards may seem to be an antiquated method of creating musical randomness in today's digitally oriented culture. *Shuffle,* interestingly, sidesteps the most familiar understanding of its title in reference to music – the shuffle option on CD players, iPods and streaming services. When the shuffle function emerged with CD players in the 1980s, it disrupted the linear order of progression determined by the artist/producer. To music critic and curator John Corbett, the shuffle feature appeared as a 'miraculous button' that undermined the 'sequential logic' dictated by the music industry and vinyl records. It ushered in an era of what Corbett terms 'musical postmodernity', whereby listeners were granted a new form of agency and meaning-making in which randomness generated not only 'unforeseeable combinations' but also 'the acceptance of incompatibility [and] the irreducibility of all forms of discourse to the logic of one'.[9] In addition to liberating listening and multiplying the options for music consumption, the act of shuffling can create its own sounds. For instance, the internal components of CD players emit whirring and clicking when shuffling between tracks and searching for data. Such shuffling sounds have piqued the interest of contemporary musicians, who have even turned the CD player into an unwitting musical instrument and compositional device. Sampling the mechanism's noises, playing cracked CDs, and ingeniously outmanoeuvring the player's optical pickup, all have foregrounded the process of shuffling in experimental music.[10] For Marclay's *Shuffle,* however, the use of an analogue medium like playing cards, where the handling of paper scraping and sliding off other pieces of paper, poses a sly counterstrategy to the digital.

Photography is key for *Shuffle*'s mobilization of affect. Marclay's snapshot aesthetic captures musical forms chanced upon while travelling in streets, stores and buildings or in observing consumer goods and commercial displays. Taken over many years, his photos reveal how the world resounds with musical graphics. Like a situationist on a dérive to discover musical psychogeographies, the artist documents the symbols of music appearing in public spaces and vernacular culture.[11] Places such as concert venues, music stores and restaurants may be expected to promote themselves with musically oriented signage and murals. The deck of photos also includes clefs, notes, signatures and staves that appear on such items as a bedspread, cookie tin, welcome mat, suspenders, sewer cover, deodorant package and toilet paper. An attitude of serendipity and humour pervades the images. The uncomposed feel relates to the quickness and sketch-like aspect of Marclay's photographic practice. He recounts that he often photographs without having a specific goal: 'I often see things after the fact . . . there's a revelatory quality . . . a sense of playfulness, because [I'm] not sure what the consequences are going to be.'[12] The creative generation of the images themselves involves trusting the affective force of gut instinct and intuition within a process of improvisational and aleatory decision-making.

The casual nature of the images belies their significance. Oddly, despite the photographs' urban milieu, the compositions are practically devoid of people, with the exception of a few close-ups showing a sequence of tattooed notes encircling an arm, a nose and chin revealed under a musically themed umbrella, or the back of a woman wearing a note-festooned shirt. The cards avoid the most obvious communicators of facial affect that might overdetermine the reading of an emotional state or psychological disposition. Indeed, the photos circumvent purposefulness by courting a prosaic affective tenor. Marclay confides that the photos were often preceded by hesitation and a questioning of the scene's worthiness: 'Most of my pictures are really small statements. There's a banality to them.'[13] Such seeming banality is far from inconsequential, however, for it strategically unfetters the image from any singular implication. Even though the emphasis on the ordinary undermines the expectation of a formally composed score, the artist's intention reveals music's presence in the vague micro-affects of commonplace scenes. And while individual musical phrases may be 'small statements' in themselves, when juxtaposed with others in a deck, such an aggregation sharpens to become compelling and energizing.

The musical notation depicted in the cards, often rendered by anonymous designers, is presented with diverse degrees of legibility. They encompass musical symbolization that is stitched, sculpted, drawn, painted, cast, inflated, printed, welded, tattooed, stencilled and wrought. Notes, staves and clefs can appear singly, as one or two bars, or in a few instances as longer fragments. While they are recognizable as notation, identifying a specific composition is for the most part futile; the photos merely divulge glimpses of partial, generic references to music. As critic Brian Sholis remarks, 'some are correctly drawn and conceivably playable on instruments; others are musical nonsense, graphic placeholders connoting "music".[14] These are the musical sequences that eventual performers of *Shuffle* must respond to as they play their instruments. As the cards carry only the appearance of sonic musicality, performers must also rely on their readings of visual affect residing in images. Not being able to read or write music, Marclay's practice is distinctive for its synaesthetic and cross-modal sensibility.[15] His artistic oeuvre is filled with works that address the intersensory nature of sound and how different registers of acoustic and visual phenomena can intersect and yet challenge simple correspondences. The deck of cards may tantalize with numerous signifiers of music, yet their performance relies on what critic Susan Tallman calls the 'incommutability of sound and vision'.[16] For the artist, 'sound' is an embodied phenomenon that exceeds hearing to involve the senses more broadly in ways that incorporate the affects transmitted by gesture, presences and atmosphere.[17]

Musical negotiation of the images on *Shuffle*'s cards hinges not only on the sounds that the notations indicate, since they provide so little information anyway, but how the images express *feeling*. Where the cards inscribe fragments of notes of a discursive system, the manner of their improvisation conveys affect through the tone of the musician's instrument, the moods of their musical style, and the spirit of how they respond to the images. The playing of the images suggests instances where sound and vision constellate to generate affect. For cultural critic Richard Dyer, the impact of popular culture genres such as musicals depends upon their power to convey feeling. He writes that beyond actors and scripts, it is the expressiveness of non-verbal and non-representational signs, which include lighting, colour, texture, movement, rhythm, camerawork and so on, that communicate the utopian spirit of musicals so convincingly.[18] The affective qualities permeating musicals, such as energy, abundance, intensity, transparency and community, hint at their origin in the privations of the Great Depression to give anguished audiences an experience of how a perfect world could possibly feel.[19] The images on Marclay's cards similarly deploy visual information to transmit affect.[20] The notes and clefs adorning objects and facades readily convey 'musicality' through non-representational codes such as the bold colours of a car dealer's sign implying the good times to be had cruising with a bad-ass stereo; the graceful curls of note-adorned fabric enveloping a Venetian mask and suggesting the festivity of a masquerade; the effervescent notes and bubbles interlaced on a champagne flute beckoning an evening enjoying affluent pleasures.

Many of Marclay's photographs, however, express ambiguous and contradictory affects. Mixed feelings are signalled by indeterminate

locations, odd angles, awkward camera flashes, mysterious shadows, or other
factors that compromise their clarity. The intelligibility of musical symbols
themselves might be obscured by the impact of time and neglect apparent
in dilapidated patinas, peeling paint, graffiti or the grime of everyday life in
the metropolis.

Each card carries affects of the ambient world from which it originates:
the tension of a body in a T-shirt, the accumulated filth on a dusty awning.
Many cards portray musical notation merging with the patterns and
textures of its context. Interpretation may become vexed when tree branches
compete with a mural's rendering of a melody. Likewise, transparencies –
such as a bar of notes on a window contending with both a ghostly guitar
and a reflection of buildings across the street – offer intriguingly confused
sequences to perform.[21] As Marclay insists, visual information can express
sonic characteristics, such as rhythm, so that 'everything [shown] in *Shuffle*
is potentially music'.[22] Critic Francois Richard adds that 'you could start
playing the decorative metal musical notes attached to the metal gate, and
then go on to play the holes in the bricks, the lines of the telephone wires,
the lines of the siding on the house, the shapes of the trees'.[23] While some
cards display a discernible affect, others may be more difficult to elucidate.
Just as the cards provide scenes of feelings and atmospheres that are vague,
variable and open to one's subjective perception, so too, their aleatory
arrangement through play comprises mutable and ever-shifting affective
relationalities.

Marclay's instructions for *Shuffle* succinctly state its premise and leave the
means of the deck's use open-ended:

> This deck of cards can be used as a musical score.
> Shuffle the deck and draw your cards.
> Create a sequence using as many or as few of the cards as you wish.
> Play alone or with others.
> Invent your own rules.
> Sounds may be generated or simply imagined.[24]

Significantly, no directions are offered on how the cards should be read,
translated or musically interpreted. In a performative situation, even before
any contemplation of playing begins, sounds are generated just in the opening
of the box, in taking out the deck, and in shuffling and arranging the cards.
The finite number of cards generates nearly infinite possibilities; yet, even if
the same cards and order were drawn by two musicians, the improvisational
element and open-endedness of the images would yield vastly different
interpretations depending on the individual's sensitivity, attitude, training,

mood and intentions. For the artist, such multiplicity is a sign of the project's success.[25]

Without the direction and continuity of standard notation, *Shuffle*'s performers must draw on their intuitive and improvisational skills. Unconventional scores using graphic and pictorial elements are, in the words of sound theorist Christoph Cox, 'structurally incomplete'.[26] Such a lack of finish challenges performers to invoke their own instinctual judgement to develop and realize the music. Ironically, according to composer Cornelius Cardew, those who are 'musically educated [are] at a disadvantage' to improvising because their training has instilled a deference to written notation.[27] *Shuffle* confounds habits of reading a linear score by relying on the assembling and arranging of individual cards. Even when cards appear in a row, a noticeable gap exists between each one, a gap that must be filled in by the performer. Transitioning from one card to the next entails intuiting some kind of bridge or connection, even when the affect of individual cards reads considerably differently. In this sense, improvisation responding to the cards becomes a collaborative co-production where the agency of the performer in many ways decentres the creativity traditionally associated with the artist.

In its shift away from the discursive conventions of musical inscription, *Shuffle* draws out the incipience of improvisation: spontaneity, freedom and possibility. Performances using the deck can involve the artist or musicians selecting cards, or cards being randomly revealed during the event itself. Either way, the performances by the musicians are created in the moment, accentuating a sense of immediacy. As Marclay explains, improvisation is 'all about the present, you do it and then it's gone'.[28] The affective state of 'presentness' induced by improvisation has been described by new music composer Vijay Iyer as one in which the 'practice of perception' is actively 're-experienced' in a 'real-time interaction with . . . one's acoustic, musical-formal, cultural, embodied and situated environment'.[29] Rather than involving an unstructured process, *Shuffle*'s improvisation involves tuning into both the sonic and visual affects of the cards, the sensing of one's body and musical training, discerning the mood of the scene of the event and disposition of fellow musicians, all of which convert into the complexity of the performance.

The results, however, can be discordant. Attending one of *Shuffle*'s performances, critic Sarah Andress described hearing sounds like 'radio frequencies, revving engines, weeping humans'.[30] A tension was evident in the audience, who emanated 'a palpable reserve that never gave way to foot tapping (or any other sort of) engagement'. Further, given the randomness of the cards, 'the dissonance that characterized much of the music wasn't surprising; what was, were the occasional moments of synchronicity'.[31] She also maintained that it was difficult, if not impossible, for the audience to figure out which cards were being played: the musicians' interpretations were for the most part 'impenetrable' and 'inscrutable'.[32] For those anticipating a harmonious experience, a cacophonous outcome might be considered a disappointment. Yet, in our understanding, *Shuffle*'s musicality or amusicality remains beside the point. In witnessing the performance, it does not matter whether the audience resonates with the sonic interpretation of the cards. The significance of Marclay's project involves being able to see, hear and feel in real time the translation and emergence of cross-modal affects.

Ultimately, as much as *Shuffle* spotlights improvisation and intuition, it also comprises a social event. As Marclay notes, both card games and performing live music 'force . . . people to sit physically together in one room'.[33] The performance of *Shuffle* mobilizes affective bonds that bring people into affiliation in the context of musical experimentation.[34] Marclay's contrasting of visual/notational systems and sonic/performative ones evokes a tension that is both challenging and artful. Whereas shuffling may be considered initially as a method to randomize cards, or a prelude to the game that follows, in Marclay's work, shuffling becomes a tactic to innovatively probe the dynamics between sound, art and affect.

Notes

1 'Shuffle', *Lexico*, https://www.lexico.com/definition/shuffle (accessed 19 August 2021).
2 On articulation, see Lawrence Grossberg, *We Gotta Get Out of This Place: Popular Conservatism and Postmodern Culture* (New York and London: Routledge, 1992).
3 Laura A. Smoller, 'Playing Cards and Popular Culture in Sixteenth-Century Nuremberg', *The Sixteenth Century Journal* 17, no. 2 (1986): 183–214. Other decks in this vein depict songs with caricatures of musicians and dancers. See Bertha Harrison, 'Games of Music', *The Musical Times* 48, no. 775 (1907): 589–92.
4 See Steve Newman, 'The Value of "Nothing": Ballads in "The Beggar's Opera"', *The Eighteenth Century* 45, no. 3 (2004): 265–83. Similarly, decks were produced to correlate music and dances, such as cotillons and polkas in the eighteenth and nineteenth centuries. See William B. Keller, *A Catalogue of the Cary Collection of Playing Cards in the Yale University Library*, Vol. 1 (New Haven: Yale University Library, 1981), 270, 274, and Felix Alfaro Fournier, *Playing Cards* (Vitoria: Fournier Museum, 1982), 219.
5 See, e.g., Keller, *A Catalogue*, Vol. II, 139–40.

6 Cage used tarot to determine the timing for *4'33"* (1952), see Kyle Gann, *No Such Thing as Silence: John Cage's 4'33"* (New Haven: Yale University Press, 2010), 175–6. Cards have also been utilized by writers, choreographers, performance artists and theatre directors who share an interest in chance juxtapositions and improvisation.

7 See also Nicole L. Carroll, 'Everything in Its Place: A Conceptual Framework for Anti- Music', *Sound Scripts* 6, no. 1 (2019): Article 15.

8 Frances Richard, 'Music I've Seen: Christian Marclay', *Aperture* 212 (2013): 28. The photos were begun in the 1990s. Similar projects of the artist's 'found sound' include *Graffiti Composition* (2002), *Zoom Zoom* (2007) and *Ephemera* (2009).

9 Corbett also recognizes that most consumers listen to a CD in the normal order and admits that shuffling pre-existing tracks can merely embody an 'empty form of eclecticism'. John Corbett, *Extended Play: Sounding Off from John Cage to Dr. Funkenstein* (Durham, NC: Duke University Press, 1994), 1.

10 See Caleb Stuart, 'Damaged Sound: Glitching and Skipping Compact Discs in the Audio of Yasunao Tone, Nicolas Collins and Oval', *Leonardo Music Journal* 13 (2003): 47–52 and Volker Straebel, 'From Reproduction to Performance: Media-Specific Music for Compact Disc', *Leonardo Music Journal* 19 (2009): 25–6.

11 See Guy Debord, 'Introduction to a Critique of Urban Geography', in *Situationist International Anthology*, ed. and trans. Ken Knabb (Berkeley: Bureau of Public Secrets, 1981), 5–8.

12 Marclay in Richard, 'Music I've Seen', 28, 32.

13 Marclay in Richard, 'Music I've Seen', 32.

14 Brian Sholis, 'Shuffle', *Print* 61, no. 4 (2007): 101. See also Christoph Cox, *Sonic Flux: Sound, Art and Metaphysics* (Chicago: The University of Chicago Press, 2018).

15 See David Briers, 'Huddersfield Contemporary Music Festival: Christian Marclay', *Art Monthly*, February 2019, 40 and Sholis, 'Shuffle', 101.

16 Susan Tallman, 'To the Last Syllable of Recorded Time: Christian Marclay', *Art in Print* 6, no. 4 (2016): 10.

17 See Teresa Brennan, *The Transmission of Affect* (Ithaca: Cornell University Press, 2004).

18 Richard Dyer, *Only Entertainment* (New York and London: Routledge, 1992), 18–19.

19 Dyer, *Only Entertainment*, 18.

20 The use of sound affects outlined here is different from what Augoyard and Torgue call 'sonic effects', which are actual sounds that have physical and psychological effects on listeners. For *Shuffle*, the affects are suggestive, intersensory and multivalent. See Jean-Francois Augoyard and Henry Torgue, eds., *Sonic Experience: A Guide to Everyday Sounds* (Montreal: McGill-Queens University Press, 2005).

21 See Kathleen Stewart, *Ordinary Affects* (Durham, NC: Duke University Press, 2007).

22 Marclay in Richard, 'Music I've Seen', 33.

23 Richard, 'Music I've Seen', 33.

24 The instructions are located on the back of the box containing the cards.

25 Marclay in Richard, 'Music I've Seen', 31.

26 Cox, *Sonic Flux*, 70.

27 Cornelius Cardew, 'Towards an Ethic of Improvisation', from *Treatise Handbook* (London: Edition Peters, 1971), Ubuweb, https://www.ubu.com/papers/cardew_ethics.html (accessed 21 August 2021).

28 Marclay quoted in Steve Tromans, 'Improvising Musical Experience: The Eternal Ex-temporization of Music Made Live', in *Experiencing Liveness in Contemporary Performance: Interdisciplinary Perspectives*, ed. Matthew Reason and Anja Molle Lindelof (New York and London: Routledge, 2016), 180. See also Cardew; and Derek Bailey, *Musical Improvisation* (Englewood Cliffs, NJ: Prentice Hall, 1980).

29 Vijay Iyer, 'On Improvisation, Temporality and Embodied Experience', in *Sound Unbound: Sampling Digital Music and Culture*, ed. Paul D. Miller (Cambridge, MA: MIT Press, 2008), 273, 280.

30 Sarah Andress, 'Musical Cards: Christian Marclay's *Shuffle*', *Art on Paper* 12, no. 2 (2007): 24.

31 Andress, 'Musical Cards', 24.

32 Andress, 'Musical Cards', 24.

33 Marclay in Richard, 'Music I've Seen', 29.

34 On affect and social bonding, see Grossberg, *We Gotta Get Out of This Place*.

Entry 3

The oriental riff

Runchao Liu

With over a century's presence in Western popular music history, the oriental riff is firmly established as a musical trope with cross-genre, transhistorical elasticity for anything related to East Asia. There have been many notable examples, including in Tin Pan Alley classic 'Chinatown, My Chinatown' (Jean Schwartz and William Jerome, 1910), the novelty trio Gaylords's 'Chow Mein' (1955), Buck Owen's chart-topping country single 'Made in Japan' (1972), Carl Douglas's mega disco hit 'Kung Fu Fighting' (1974), New York Dolls's punk rendition of 'Bad Detective' (1974; a cover of the Coasters's 1964 single of the same title), progressive rock band Rush's 'A Passage to Bangkok' (1976), the Vapors's new wave one-hit wonder 'Turning Japanese' (1980) and Day Above Ground's notorious 'Asian Girlz' (2013). This musical trope may have an origin as early as nineteenth-century fantastical musical plays and has been popularized throughout Western mediascapes, including but not limited to operas, films, video games and cartoons.[1] The oriental riff (alternatively, the Asian riff, Chinese riff or Asian jingle) has become a formulaic sound of exoticism. While musicologists have produced much scholarship on musical exoticism and orientalist musical devices in Western art music, little attention is paid to contemporary popular music with a few exceptions focusing on early popular music.[2] It is this research gap in relation to the sound-affect relationship that I seek to address in this entry.

Repetition is of key relevance here in terms of both musical characters and Asian racialization. The oriental riff is a cliché melody that most typically features staccato articulation and pitch repetition with occasional pentatonic hints, often used in the opening measures of a song or between the lyrics as musical ornaments to signal and construct an affectively exotic soundscape. It can vary in terms of tonality, timbre, harmonization (parallel fourths, fifths and octaves are the most common) and the exact number of notes. In contemporary popular music, Jamaican singer-songwriter Douglas's 1974 disco hit 'Kung Fu Fighting' axiomatically represents the revival of the oriental riff in popular music with its international commercial success.

Figure 3.1 The oriental riff used in Carl Douglas's 'Kung Fu Fighting' (1974), transcribed from 01:07 to 01:09. Transcription by the author.

It also popularized a quintessential version of the oriental riff (see Figure 3.1) that contemporary audience would find most familiar: a melody that consists of nine notes stretched out over two 2/4 or 4/4 bars and plays with three pitches. Many musical employments of the oriental riff can be simply transposed from this version, such as in 'Turning Japanese' and 'Young Folks' (Peter Bjorn and John, 2006).

Both sounds (aural stereotyping) and affects (character stereotyping) have been integral to the racial stigmatization and Western imagination of Asians. From heavy accents to sound-based racial slurs like 'Ching Chong', crashing gongs in films to cue the entrance of a person of Asian descent and the oriental riff, the aural has always been a source of and witness to Asian racialization. Aural stereotyping like 'sounding Asian' and 'yellowvoice' is no less detrimental than slanting eyes or waving karate fists.[3] At the same time, the stereotyping discourse of unfeeling Asians has been established in as early as the fourth and fifth century BCE in the Greek philosophical representation of Asians as monotonous, emotionally stagnant and lacking spirit through the words given to them by Hippocrates and Aristotle.[4] The racial trope of the taciturn, robotic and unfeeling foreigner was revived in popular culture as a techno-orientalist discourse after East Asia's rapid economic growth in the 1980s.[5] Given such a historical background, the emergence and undying popularity of the oriental riff are not without reason. It functions both sonically and affectively through a kind of multisensory performativity, wielding its reiterative power to constitute, reproduce and consolidate what Jennifer Lynn Stoever calls 'the sonic color line'.[6] Considering the oriental riff's generic affective power of exoticism and sonic feature of rhythmic and pitch repetitiveness, it is an 'ideal' object for us to examine the sound-affect relationship through the lenses of race, gender and sexuality.

By approaching sounds and affects as mutually constitutive and constantly mediated, this entry develops the concept of sound affects as an intersectional framework to broaden our ways to examine processes of racial formation beyond the tendency to centralize the optic. While many have examined how musical co-optation and racially charged sounds bring affects to serve and reflect the image of their Western creators, it is often ignored how affects also influence sounds and how sounds and affects are constitutive of each other. It should not come as a surprise that I need to stress that the sound affects of the oriental riff I examine in this entry, including exoticism, mysticism, empowerment, pseudo-progressiveness and eroticism, are not any 'new' orientalist creations. Instead, what I argue is that while the oriental riff has produced a variety of intersected and sometimes contradictory affects, these well-established sensibilities associated with the exotic orient have also distinctively inflicted the ways the oriental riff and other Asian elements sound and perform in music. In this way, not only do racialized sounds produce exotic affects, these affects also influence and operate as sounds. Due to the imbricated and contradictory trajectories of racialized sonic signifiers and racialized affects, the oriental riff encourages us to rethink the concept of sound affects as a mutually constitutive relationship between sounds and affects, which we may witness through dissecting processes and challenges of Asian racialization manifested in some sophisticated musical examples.

Meanwhile, this entry employs the concept of sound affects to demonstrate the discursive pliability of the oriental riff, a formulaic and supposedly stable musical form. In doing so, we may interrogate sound affects as a productive decolonizing tool of listening, which exposes the constructive frailties of aural racialization. My anti-essentialist approach and emphasis on contradiction and inconsistency are in line with Christine Bacareza Balance's idea of 'disobedient listening', which turns to the sonic and musical to challenge hegemonic listening and identify the limits and demands of the cultural logic of racial visibility.[7] My examination of the oriental riff's sound affects is not only informed by the theoretical framework of disobedient listening but also demonstrates how we may apply disobedient listening as a method of critical listening to sound affects. Particularly, Balance positions such practices as both disruptive (against dominant discourses) and generative (for silenced perspectives). Listening both disruptively and generatively not only informs my examination of various adaptations of the oriental riff but also where I suggest future researchers to look to take inquiries of the oriental riff and its sound affects further. Through examining how Filipinx America practices disobedient listening, Balance argues that tropical renditions that gesture at and reimagine the Philippines in the aftermath of US imperialism emerge. To this end, practising disobedient listening to sound affects allows us to resituate the oriental riff as a floating signifier. As noted towards the end of this entry, listening to the oriental riff disobediently takes us to Asia and Asian America to further contest the meanings and sound affects of this musical trope. Before that, in what follows, I compare various appearances of the oriental riff to unpack how and why they sound affectively in various

ways through Douglas's 'Kung Fu Fighting', Rush's 'A Passage to Bangkok',
David Bowie's 'China Girl' (1983), Day Above Ground's 'Asian Girlz' and
Siouxsie and the Banshees's 'Hong Kong Garden' (1978).

Disobedient listening: Sounding and feeling the oriental riff

The oriental riff has a history of over a century and its musical origin
can be traced back to Thomas Comer (composer) and Silas S. Steele's
(lyricist) fantastical musical play *The Grand Chinese Spectacle of Aladdin
or The Wonderful Lamp* produced at the Boston Museum in 1847. This
is a stage adaptation of *Aladdin*, a popular story in the early nineteenth
century in both Great Britain and the United States. Although no libretto
of Comer and Steele's adaptation exists, we can find similar rhythmic
patterns and pitch repetition in the sheet music of at least two songs of
the play, 'Come, Come Away' and 'Aladdin Quick Step'. What now seems
like an awkward combination of the Middle East and China was in fact
popular in stage shows for its spectacle value, although the stories of
Aladdin were often set in western China. Moreover, many American
actors playing *Aladdin* were blackface minstrels who combined blackface
and yellowface traditions.[8] This orientalist custom, conflating multiple
Asian and African identities, provided the larger cultural context out of
which the oriental riff was born, reminding us from the outset of the
fictitious nature of the trope.

While commonly used to signify a generic sense of mysticism and
exoticism, the oriental riff is seldom used without being complicated by
other simultaneous, albeit often ignored, sound affects. What follows is
a comparison of Douglas's 'Kung Fu Fighting' and Rush's 'A Passage to
Bangkok' in their use of the oriental riff to illustrate how simultaneous
affects are produced. In 'Kung Fu Fighting', the oriental riff is introduced
in the opening section and played on the keyboard to emulate the timbre
of a Chinese flute, conveniently interpellating the melody into a sonic hook
to spotlight the song's subsequent worldmaking. The lyrical reference to a
surreal Chinatown where everyone is a kung fu master making lightning-
fast kicks, coupled with the frequent use of the oriental riff throughout the
chorus, confirms the adoption of this riff as a sonic tool to construct an
exotic sense of mystique. All that notwithstanding, one much undervalued
sound affect being conjured up here is the empowering affect of cross-racial
'kung fusion' – a word play of kung fu and fusion used by Vijay Prashad to

encapsulate the Afro-Asian solidarity that has blossomed around kung fu, an anti-hierarchal martial art empowering oppressed and poverty-stricken people.[9] Furthermore, the notion that everyone can be a kung fu master, that is, one's own hero, reverberates throughout the song and lends itself effortlessly to the much-needed solidarity of 'kung fusion', despite a cheesy one. A distinct musical trope, the oriental riff plays a critical role in ensuring the delivery of this empowering affect.

In comparison, the oriental riff used in Canadian heavy metal band Rush's 'A Passage to Bangkok' also constructs nuanced affects. It is conspicuous in the lyrics that this passage to Bangkok is a cross-continental drug tour that passes the Caribbean Sea, Mediterranean Sea, the Middle East, South Asia and Southeast Asia. Employed in the opening measures and delivered through the heavily distorted tone of electric guitar, the oriental riff here is a welcome note that also serves to warn us about the mysterious and perhaps dangerous nature of this tour. The choice of the riff, a conspicuous East Asian orientalist musical sign, to initiate a trip that has nothing to do with East Asian cities is not unforeseen considering the epidemic association of the Chinese with pleasures of opium in popular music.[10] Doing so also taps into heavy metal's habit of appropriating Eastern themes as sources of power.[11] Moreover, the song employs a semitone (C#) in the oriental riff (D – C# – A – C#) instead of what would be a standard version (D – C – A – C) to create an unorthodox, drug-induced and racialized sense of power and to indicate the beginning of a much-anticipated exotic trip. Although like it in 'Kung Fu Fighting', the oriental riff employed in 'A Passage to Bangkok' also works at the intersection of exoticism, mysticism and empowerment, the latter works more through established connotations of musical genre and tonality and less through cross-racial solidarity. As the audience, we must attend to these different writing techniques and how rearrangements are used to construct affectively unique soundscapes. To generalize the meanings and affects of the oriental riff simply as exoticism without attending to its affective nuances would be to give in to the essentializing tendency of such racialized musical tropes.

The oriental riff is also often used to create racialized erotic sensibilities. By zooming in on the seemingly harmless representation of the nameless Asian woman in John Mellencamp's 'China Girl' (1982), Bowie's 'China Girl' and John Zorn's 'Forbidden Fruit' (1987), Ellie M. Hisama observes a neocolonialist, heterosexist 'rescue fantasy' filled with romantic desires for Asian women in the 1980s popular music that she calls 'Asiophilia', a sexual extension of James Clifford's idea of 'salvage paradigm', seeking to 'save' authentic cultures from waning.[12] A modern, glissando guitar adaptation of the oriental riff (sliding between B – A – G – A) stacked with parallel fourths and followed by xylophone-like crispy sounds announces the beginning of Bowie's 'China Girl'. The revised riff seems to imply an interracial love story that aims to, albeit fails to, divert from the pervasive fetishistic representations of Asian women. That said, what follows this 'Chop Suey guitar intro', as Karen Tongson calls it, is an 'imperial foreplay gone terribly awry'.[13] As portrayed in the music video, their story ends with the China Girl and Bowie having sex while being fully naked and gently lashed by the waves on a beach. The supposedly yet questionably progressive construct of interracial eroticism is limited by its salvage storyline inscribed in both the lyrics and music video as the narrator warns the China Girl about his courtship for what he represents – the modern, the ambitious, the whiteness – would ruin what is so precious about her. Together, the salvage theme, Bowie's outspoken anti-racism and the ambivalent autonomy to 'shut your mouth' bestowed to the China Girl add a pseudo-progressive sensibility to the erotic atmosphere. The intentionally altered yet fatally recognizable oriental riff captures the song's contradictory sensibilities of eroticism, progressiveness and heroism.

In contrast, the oriental riff in the American band Day Above Ground's pop single 'Asian Girlz' is self-evidently one of the most outrageous adoptions in recent times. The song has caused a media firestorm and has been widely criticized as racist, fetishistic and hyper-sexualizing with obnoxious lyrics that use various racially charged descriptors on Asian women, such as 'yellow' on thighs and 'slanted' on eyes. The music video features Vietnamese American actress Levy Tran, who remains nameless as the fetishized Asian Girl, stripping down in front of the band members, who later all jump into her bathtub. Just as minimum effort is used to disguise the song's Asiophilia, the oriental riff is also adopted with minimum adaptation so that nobody could mistake the sonic marker for anything else: a single unaccompanied melody (C# – B – A# – B) directly transposed from what we hear in the chorus of 'Kung Fu Fighting' and played in reverb effects to create an ethereal touch. This rather standard version of the oriental riff is used throughout the song and every single time is used to direct the ethnic gaze at Tran's spinning body in lingerie or close-up shots of her breasts and teeth licking. The sensibilities of pseudo-progressiveness that would have been possible in 'China Girl' are gone without a trace in 'Asian Girlz'. What is left instead is musical orientalism in its barest possible way.

In the shadow of the modernist ethnic gaze that treats ethnic Others as inferior and in need of salvation,[14] the postwar milieu and cross-cultural collaborations in countercultural and avant-garde circles between the 1960s and 1980s injected the ethnic gaze with a refreshing, albeit not new, sensibility of pseudo-progressiveness. In other words, each political conjuncture decides the changing meanings of its oriental Others. For instance, with regard to the representations of Japan in American popular culture, while japonisme suggested an elite status for some white Americans in the late nineteenth century amidst rapid economic growth, anti-Japanese sentiments became prevalent during the Second World War in wartime pop songs and Japanese brutality and atrocities were frequently portrayed in Hollywood and US propaganda films.[15] In the postwar era of the 1960s, popular culture became less explicitly political, providing some space for counterculture to shape its own relationship with the Other by getting involved with Eastern philosophies and cultures.[16] However, counterculture's 'newfound' musical affective alliance with the non-West was only pseudo-progressive in that it paralleled both a normative romanticizing use of the Other and the politics of US-Vietnamese relations whereby America practised a neocolonialist agenda.[17] This might be epitomized by the case of the unprecedentedly successful benefit concert, *Concert for Bangladesh*, co-organized by George Harrison and Ravi Shankar. Cross-racial, humanitarian projects like this inevitably reproduced the neocolonialist discourses of Western salvation and Eastern (in this case Bangladesh/then-East Pakistan) poverty-stricken refugees.[18] Such a political and musical trajectory lent its pseudo-progressive sensibilities to an array of practices of musical exoticism.

Another pseudo-progressive, humanitarianism-gone-wrong example is English postpunk band Siouxsie and the Banshees's 'Hong Kong Garden'. Against the backdrop of the Rock Against Racism movement in the UK, anti-racist voices were growing exponentially in all pop music genres. Although vocalist Siouxsie expressed their indifference to the movement, their 'Hong Kong Garden' has a clear anti-racist behind-the-scenes story about how Siouxsie was disgusted by the racist customers at a takeaway named Hong Kong Garden when she was growing up.[19] However, the lyrics are as careless and problematic as they can be, suggesting that Japanese currency yen is somehow used at a Chinese takeaway in southeast London while casually describing the workers there having 'slanted eyes'. In this ambivalent anti-racist song, a line of notes curiously resembling the oriental riff begins the song in staccato with xylophone-like timbre (see Figure 3.2), which is recognized as an 'echoic "oriental" guitar figure' in a 1978 review in *Sounds*.[20] While the opening melody might not instantly conjure up a quintessential oriental riff, with a little rearrangement we can see that the melody can be easily converted into a quintessential oriental riff using the exact same three pitches (see Figure 3.3). What makes this formulation of the oriental riff distinct from many others, though, is its more dominant role in the song. The melody of the riff is also the base melody of the verses, which is one of the rare occasions where the riff assumes more of a constitutive role

Figure 3.2 The opening measures of Siouxsie and the Banshees's 'Hong Kong Garden' (1978), transcribed from 00:00 to 00:03. Transcription by the author.

Figure 3.3 The conversion of Siouxsie and the Banshees's 'Hong Kong Garden' (1978) into a more standard oriental riff. Conversion by the author.

than solely functioning as decorative insertions. This musical approach, as some have argued, can still be problematic because it exemplifies a kind of neocolonization.[21]

From progressive rock to postpunk, the oriental riff has witnessed this 'newfound' affective alliance with East Asian and to a lesser degree Southeast Asian peoples through its various appearances. By virtue of such political and artistic legacies, it is expected to find critics like Charles Shaar Murray in a 1984 issue of *New Musical Express* applauding Bowie for his controversial 'China Girl' by pointing out the song's political nature and citing the general prejudice against pop stars' political interventions in defence of Bowie, writing that 'the political activities and statements of pop stars are generally devalued by the general received notion of entertainers as privileged cranks and eccentrics'.[22] Some may notice the parallel between the hyperbolic portrayal of the stereotypical Chinese woman and the parody of the oriental riff to suggest that the song's overall artificiality as to representation could all be Bowie's creative devices for articulating a genuine interracial relationship.[23] The pseudo-communist-soldier persona of the China Girl in the music video also projects some political progressiveness in a historical moment still fraught with Cold War tensions. However, just as Murray suggests, 'China Girl' is Bowie's political statement, which, as Hisama observes, is at the expense of the pursued China Girl and relies on an 'implicit hierarchical nature of otherness [which] invites seemingly innocuous practices of representation that amount to (often unknowing) strategies of domination through appropriation.[24] In short, a disobedient listening through a multimodal discursive approach to 'China Girl' reveals not only the mutually constitutive relationship of its sounds (the rearranged oriental riff) and affects (ambiguously progressive erotic sensibilities) but also how unintended affects (neocolonialist sensibilities) continue to rewrite the meanings of the original sounds long after the completion of a song or a performance.

Final notes

Exotic musical gestures invite the ethnic gaze, aestheticizing processes of Asian racialization and ethnic differences. As shown earlier, the adoptions and variations of the oriental riff in popular music abound. Different from raga rock and the 'turn to the East' in the 1960s, which to a great extent was concerned with exploring South Asian and North African musics, spiritual search and producing authentic experiences,[25] the turn to the Far East in the 1970s and 1980s has been markedly more superficial and commercially minded. Musical exoticism, however, with its long history of cultural confusion, is a question more about its function than its authenticity.[26] This way, the oriental riff and an actual gong sound do not make much difference as exotic signifiers. Therefore, we may better approach the sonic constructs of the exotic as sound affects through which we shall place our focus on the shifting, ambivalent and generative affective power and experiences associated with the function of the oriental riff across time and space.

Through disobedient listening, my examination complicates the sound-affect relationship of the oriental riff and reveals its internal contradictoriness as an exotifying sonic phrase. Focusing on inconsistency and contradiction reveals the interstices of the sonic phrase's signifying practices and is more likely to avoid replicating orientalist discourses while studying them. Thus, disobedient listening also leads us to better address the disobedient playing of the oriental riff by those who are often its signified exotic bodies. For example, the original punk soundtrack to the experimental comedic short film *Shut Up White Boy* (directed by Vũ T. Thu Hà) prominently reappropriates the oriental riff to signpost the narrative diversion to satirizing Asiophilia through a series of creative revenge acts on an irritating white boy with 'yellow fever'.[27] Biracial Korean American musician Michelle Zauner's indie single 'Everybody Wants to Love You' (2016) features a playful intro riff that hauntingly reminds us of the commonplace use of the oriental riff in popular music. Korean American comedian Margaret Cho also calls out how the oriental riff and other racist routines were used in a college cheerleading tournament.[28] There are also cases where the oriental riff is not used in explicitly counter-hegemonic ways and sometimes predominantly circulated in East Asia. For the Chinese audience, the oriental riff is perhaps more well known as a video game melody and still not widely recognized. The commercial success of the *Kung Fu Panda* franchise along with its revival of 'Kung Fu Fighting', including its Chinese version, has helped make the melody more familiar to the Chinese audience. In Taiwanese popular music, there are a few notable examples that sampled the melody, including MC Hotdog's 'Han Liu Lai Xi' (trans. 'Korean Wave Invasion', 2001), Shan-Wei Chang's 'He Quan' (trans. 'Crane Style Boxing', 2003) and Cyndi Wang's 'Jie Mao Wan Wan' (trans. 'Curved Eyelashes', 2005). Another one is 'Xing Zuo' (trans. 'Astrology', 2004) by Leehom Wang, an American musician but also mega pop star in East Asia, who is known for his musicianship of fusing Western popular genres with traditional Chinese music. All these cases would require further examination in their respective contexts. Elastically employed at the intersections of various affective tendencies and intensities across time, genres and geopolitical contexts, the oriental riff transcends the boundaries of sonic ethnic chic/kitsch/camp and continues to offer transhistorical, transcultural and transnational insights into the changing relationship of sounds and affects.

Notes

1   Krystyn R. Moon, *Yellowface: Creating the Chinese in American Popular Music and Performance, 1850s–1920s* (New Brunswick: Rutgers University Press, 2005); Thomas Solomon, 'Music and Race in American Cartoons: Multimedia, Subject Position, and the Racial Imagination', in *Music and Minorities from Around the World: Research, Documentation and Interdisciplinary Study*, ed. Ursula Hemetek, Essica Marks, and Adelaida Reyes (Newcastle: Cambridge Scholars Publishing, 2014).

2   Ellie M. Hisama, 'Postcolonialism on the Make: The Music of John Mellencamp, David Bowie and John Zorn', *Popular Music* 12, no. 2 (1993): 91–104; Judy Tsou, 'Gendering Race: Stereotypes of Chinese Americans in Popular Sheet Music', *Repercussions* 6, no. 2 (1997): 25–62; Charles Hiroshi Garrett, 'Chinatown, Whose Chinatown? Defining America's Borders with Musical Orientalism', *Journal of the American Musicological Society* 57, no. 1 (2004): 119–73.

3   For the idea of 'yellowvoice', see Robert Charles Lancefield, 'Hearing Orientality in (White) America, 1900–1930' (PhD diss., Wesleyan University, 2004).

4 Gary Y. Okihiro, 'When and Where I Enter', in *Asian American Studies Now*, ed. Jean Yu- wen Shen Wu and Thomas C. Chen (New Brunswick: Rutgers University Press, 2010), 3–20.

5 David S. Roh, Betsy Huang, and Greta A. Niu, 'Technologizing Orientalism: An Introduction', in *Techno-Orientalism: Imagining Asia in Speculative Fiction, History, and Media*, ed. David S. Roh, Betsy Huang, and Greta A. Niu (New Brunswick: Rutgers University Press, 2015), 1–19.

6 Jennifer Lynn Stoever, *The Sonic Color Line: Race and the Cultural Politics of Listening* (New York: New York University Press, 2016).

7 Christine Bacareza Balance, *Tropical Renditions: Making Musical Scenes in Filipino America* (Durham, NC: Duke University Press, 2016).

8 Moon, *Yellowface*, 24.

9 Vijay Prashad, *Everybody Was Kung Fu Fighting: Afro-Asian Connections and the Myth of Cultural Purity* (Boston, MA: Beacon Press, 2001).

10 Moon, *Yellowface*, 121–2.

11 Robert Walser, *Running with the Devil: Power, Gender, and Madness in Heavy Metal Music* (Hanover: University Press of New England, 1993), 154.

12 Hisama, 'Postcolonialism on the Make', 95; James Clifford, 'Of Other Peoples: Beyond the "Salvage" Paradigm', in *Discussions in Contemporary Culture #1*, ed. Hal Foster (Seattle: Bay Press, 1987), 121–30.

13 Karen Tongson, 'China Girl', in *Black Star Rising and the Purple Reign*, ed. Daphne Brooks (Durham, NC: Duke University Press, Forthcoming).

14 Meenakshi Gigi Durham, 'Displaced Persons: Symbols of South Asian Femininity and the Returned Gaze in U.S. Media Culture', *Communication Theory* 11, no. 2 (2001): 201–17; George Lipsitz, *Dangerous Crossroads: Popular Music, Postmodernism, and the Poetics of Place* (London: Verso, 1994), 4–5.

15 W. Anthony Sheppard, 'An Exotic Enemy: Anti-Japanese Musical Propaganda in World War II Hollywood', *Journal of the American Musicological Society* 54, no. 2 (2001): 303–57; W. Anthony Sheppard, *Extreme Exoticism: Japan in the American Musical Imagination* (Oxford: Oxford University Press, 2019), 54.

16 Although the oriental riff has a limited presence in the 1960s, this decade witnessed an unprecedented popular musical interest in Indian and South Asian musics. See Jonathan Bellman, 'Indian Resonances in the British Invasion, 1965-1968', *The Journal of Musicology* 15, no. 1 (1997): 116–36; Brian Ireland and Sharif Gemie, 'Raga Rock: Popular Music and the Turn to the East in the 1960s', *Journal of American Studies* 53, no. 1 (2019): 57–94.

17 Nadya Zimmerman, *Counterculture Kaleidoscope: Musical and Cultural Perspectives on Late Sixties San Francisco* (Ann Arbor: University of Michigan Press, 2008), 62–4.

18 Nayanika Mookherjee, 'Mobilising Images: Encounters of "Forced" Migrants and the Bangladesh War of 1971', *Mobilities* 6, no. 3 (2011): 399–414.

19 Carol Clerk, 'The Making of "Hong Kong Garden"', *Uncut*, May 2008, https://www.rocksbackpages.com/Library/Article/the-making-of-hong -kong-garden (accessed 1 March 2021).

20 Peter Silverton, 'Siouxsie & the Banshees: The Scream', *Sounds*, 14 October 1978, https://www.rocksbackpages.com/Library/Article/siouxsie--the-ban-shees-the-scream (accessed 1 March 2021).

21 Timothy Dean Taylor, *Global Pop: World Music, World Markets* (New York: Routledge, 1997), 41–4; Susan Fast, *In the Houses of the Holy: Led Zeppe-lin and the Power of Rock Music* (Oxford: Oxford University Press, 2001), 101–2.

22 Charles Shaar Murray, 'David Bowie: Sermon from the Savoy', *New Musical Express*, 29 September 1984, http://www.rocksbackpages.com/Library/ Article/david-bowie-sermon-from-the-savoy (accessed 1 March 2021).

23 Shelton Waldrep, 'The "China Girl" Problem: Reconsidering David Bowie in the 1980s', in *David Bowie: Critical Perspectives*, ed. Eoin Devereux, Aileen Dillane, and Martin J. Power (New York: Routledge, 2015), 147–59.

24 Hisama, 'Postcolonialism on the Make', 91.

25 Susan McClary, *Conventional Wisdom: The Content of Musical Form* (Berke-ley: University of California Press, 2000); Ireland and Gemie, 'Raga Rock'.

26 Carl Dahlhaus, *Nineteenth-Century Music*, trans. J. Bradford Robinson (Berkeley: University of California Press, 1989), 302; Sheppard, *Extreme Exoticism*; Derek B. Scott, 'Orientalism and Musical Style', *The Musical Quarterly* 82, no. 2 (1998): 309–35.

27 The soundtrack is created by a punk band under the name of the Drag-On Ladies, which caricatures the Asian dragon lady stereotype and plays the drag of queer drag culture.

28 This can be found in Cho's 1996 comedy CD *Drunk with Power*. The refer-ence is at the end of the track 'Asian-Americans – Racists'.

Part II

Voices and vocals

aa ee ii oo uuuuu

Rachel Shearer

This entry explores the vowel as a sound affect through the *kōauau* – a flute from *te ao Māori* (the Māori world), emphasizing several traditional concepts within this worldview. Here, the *kōauau* is an extended family member with agency and a 'voice'. The *kōauau* is part of a vast genealogical network that links the human to the non-human, traversing generations back to and beyond the existence of matter. Aspects of this worldview may be familiar through posthuman theories which recognize the agency of 'things' in an interconnected material world, but there are specifics to the Māori world where these networks resonate with life energies and stories that simultaneously bind the physical to the metaphysical, resonating through cosmogonies, time and space. Here the *kōauau*'s 'voice' emotes and projects the open vibratory vowel, opening portals to a myriad of potential affects in *te ao Māori*. Affect is understood here as the successive energetic exchanges of life force between entities in response to an event in an interconnected universe of sensitive/sensing webs of vibrant matter.

Whakapapa

In building this discussion and its contexts, we start from the ground. As academic Carl Te Hira Mika explains in the context of research: "'Ground" in a Western sense can be thought of as the metaphysics that you base a perception on. This metaphysics relates to finality, solidity of foundation, and an ultimate set of rules or categories that can be applied to things.'[1] He goes on to explain that the 'ground' in *te ao Māori* is inextricably linked with the idea of simultaneously being of, in and on the earth. As Māori exist within Papatūānuku (the earth personified as primordial mother), 'she infiltrates everything we do, including thinking'.[2] This idea is embodied in the word *whakapapa*, a word that is both physically and conceptually immediate and immense. The prefix *whaka* refers to causing something to happen, to become like, to move towards. The *papa* in *whakapapa* relates to something broad or flat, and also, according to musician and educator Te Ahukaramū Charles Royal, *papa* refers to Papatūānuku.[3] Entities, forces, things are personified in the Māori world.

Personifications depict the natural world as ancestors and extended family members. Therefore, as Mika has previously suggested in a glossary of Māori terms, *whakapapa* could be understood in one sense to 'become (move towards) earth'.[4] *Whakapapa*, as a verb, is to physically layer one thing on top of another. The stratification implied here is also in the use of the word as it pertains to a method of organizing histories – recounted as genealogies that structure the nature of the universe and knowledge itself. The practice of reciting these genealogies and the stories they contain, histories of human and non-human, material and immaterial forces, are performed in the correct order through rich oral traditions. *Whakapapa* is the map of the interconnected webs of the material and immaterial forces through space and time. The interwoven sequential histories within Papatūānuku and their acoustic emissions provide the 'ground' for this entry.

Kōauau

In the creation *whakapapa* of the physical world, the primordial parents Papatūānuku, the earth as mother, and Ranginui, the sky/heavens as father, begat multiple children who are the *atua* (ancestors/divine presences), personifications of and rulers of the different natural elements. In one widely held version, their son Tāne, in his various forms, is the ancestor of humans, forests, birds, insects and ultimately the *kōauau*. *Taonga puoro* (literally singing treasures), the Māori traditional instruments, are grouped in families descended from Papatūānuku and Ranginui. For example, the bellow of conch shell trumpets is given voice by their ancestor Tangaroa, *atua* of the sea. The spinning *pūrerehua*, also known as a bullroarer, an instrument made of wood, stone or bone attached to a long string, descends from Tawhirimātea, *atua* of the winds and air, who give it its whirring tonal breath.[5] Tāwhirimātea is also the carrier of sound and with it the *wairua* (spirit) of the sound carried, providing a vehicle for sound communication.[6] It is worth highlighting that the creation stories of *te ao Māori* can equally be symbolic or allegorical references to scientific fact and the energetic/material relationships between entities.

The *kōauau* is from the flute family of Raukatauri, the *atua* of music and personification of the common bag or case moth (liothula-omnivora), endemic to Aotearoa New Zealand. She makes an appearance in origin stories of performance and entertainment but the most recounted story is of how

Raukatauri loved her flute so much that she changed herself into the bag moth so she would live inside her flute-shaped cocoon and never be parted from it. Inside her cocoon she is impregnated, hatches her eggs, her sons and lovers always leave her and fly away, she has no wings. Her songs are a mournful lament for them. The female bag moth doesn't literally have a voice we can hear.

The family of flutes are the largest group of *taonga puoro*, with *kōauau* the most common. They were traditionally made from wood, kelp and bones of both animals and humans.[7] Like the human larynx, the *kōauau* is essentially a hollow tube that opens at both ends. Generally, they have three finger holes on the top and one at the back. The end at which you press your mouth to play is known as the *waha* or *māngai* (mouth). The *kōauau*, like many musical instruments, can be a conduit for expressing emotion but like all *taonga puoro*, the *kōauau* is more than just a musical instrument. The *kōauau* are also tools that enable a player to commune with the spiritual realm, they are devices that can channel the 'voices' of the *atua* who have influences over particular domains and they are also an extension of the human voice.[8] Each instrument possesses individual qualities, and they are given names and understood as 'things' with their own *mauri* (life force).

The *kōauau* pictured in this entry is named Ngōiro after a story about my ancestors, Hekeiterangi and Mokaiohungia, from around the seventeenth century that is recounted in Jo'el Komene's research on *kōauau*.[9] Hekeiterangi was charmed into marriage by Mokaiohungia, a 'virtuoso of the Māori flute' from Rongowhakaata *iwi* (extended kinship group).[10] In Leo Fowler's records, Waioeka Brown explains that Mokaiohungia caught a *ngōiro*, a southern conger eel (Conger verreauxi), 'and cooked it in a *hangi* (earth oven)'.[11] Mokaiohungia's friend, Mokaiiwike, was going to pay him a visit, and his wife suggested that they eat the eel before he arrived. After hearing of this, Mokaiiwike, unimpressed and in response, encouraged Mokaiohungia to marry Hekeiterangi (even though both were already married). Mokaiohungia heeded his friend's advice and 'played his flute to such effect that he won the lady's affections'.[12]

In Māori oral literature many stories confirm that skilled flute-playing made the performer highly attractive. One well-known story is the romance of Hinemoa and Tūtānakei. They were forbidden to be together since Tūtānakei, although the son of a chiefly family, was illegitimate, and for that reason was not an equal to the high-born Hinemoa. Tūtānekai played beautiful music every evening with his best friend Tiki, who may have also been his lover but that is another story.[13] The sound of the flute is said to have guided Hinemoa to swim the three-kilometre distance across Lake Rotorua to successfully unite with him on Mokoia Island. However, there are other versions, such as the one told to Richard Nunns by elders of Te Arawa *iwi* from where the story originates, which claim that in fact Tūtānekai couldn't play the flute well and that the real talent was someone crouching in the bushes, maybe Tiki, while Tūtānekai mimed.[14] Suggestions of trickery sometimes emerge in the seductive affects of the *kōauau*.

Stories that exist in Māori oral literature offer some insight into how these instruments were regarded. Many stories, practices and knowledge around *taonga puoro* have been lost due to the degradation of, and alienation from, Māori cultural identity that occurred during colonization. By the 1980s, *kōauau* were largely only seen in museums and rarely heard. The group Haumanu led by renowned Tūhoe composer Hirini Melbourne, working with Nunns, Brian Flintloff and many more, led the initiative to revive the knowledge and use of these instruments.[15] It is through their workshops within Māori communities that stories and memories were revived, shared and collected. In Nunn's book that records some of their research, Napi Waaka from Te Arawa *iwi* recalls sitting under a table as a small child listening to their elders discuss how to play the *kōauau* to make a baby. He also describes how *kōauau* were frequently played at *tangihanga* (funerals) to express grief and sorrow. There was a specific method of playing the *kōauau* that he had heard called *tuamatangi* – described as the respiration before death, or the dying gasp.[16] *Kōauau* were also used to aid in healing and pain relief. They were played during the application of *tā moko* (traditional tattooing). An unnamed informant from Te Whānau-a-Apanui *iwi* reported lengthy and regular playing of *kōauau* over broken bones to aid healing.[17]

In an episode of the Māori documentary programme *Waka Huia*, Hirini Melbourne states: '*Kei tēnā rākau, tōna reo, tōna reo, tōna reo*: each type of wood has its own voice.'[18] The *kōauau* I was gifted is carved from Tōtara (Podocarpus totara), a softer wood that creates a softer sound as the body of the *kōauau* will 'vibrate and release some of the sound through the wall of the instrument', as opposed to a *kōauau* made from, for example, Pūriri (Vitex lucens), which produces a crisper sound due to the hardness of the wood.[19] Ngōiro, the *kōauau* that is the 'object' of this entry, was carved by Tim Codyre and gifted to me by him and his partner Waimania Wallace of Tūhoe *iwi* (see Plate 2). Codyre was trained and mentored by Alan Nopera, a renowned carver from Ngāti Whātua *iwi*. Different *iwi* and *rohe* (districts) have different styles of carving.

Mauri

Before we move into discussing the sound affect *aa ee ii oo uuuuu*, we need to establish another aspect of our 'ground'. Where *whakapapa* gives us structure, a map, a 'taxonomy of the universe' we need to introduce the energies and forces that are immanent to the Māori world.[20] If we consider vitalism as a theory that understands that the origin and phenomena of life are dependent on forces different to those that are chemical or physical, a Māori vitalism can be located in the belief that an immanent life force 'imbues and animates all forms and things of the cosmos'.[21] As Te Kawehau Hoskins and Alison Jones citing Māori academic Mason Durie explain: 'Within a Māori ontological frame, all beings and objects are experienced as having mana, a form of presence and authority, and a "vigour, impetus, and potentiality" called mauri'.[22] The life forces known as *mauri* and *mana* exist within an intricate relationship of metaphysical concepts – *tapu, wairua* and *hau* – that together inform the expression of a Māori vitalism. These terms are difficult to do justice to in the space of this entry. There is no 'one' word in English that will translate them or communicate their nuanced roles in combination with each other, their rich contextual meanings, their 'metaphorical fluidity and their multi-layered ability to access the real'.[23] However, a simple translation introduces *tapu* as meaning sacred, prohibited, restricted, under *atua*/divine protection. *Wairua* is a spiritual essence that enters the body at embryo stage and exists after the body has died,[24] but in different contexts, and in this entry, it is used to describe the mood, atmosphere or essential attitude of something. *Hau* is closely entwined with the energetic essence of *mauri*. It is also a vital energetic essence of a person, place or object. Its vitality attracts the vitality of other beings – for example, the *hau* of trees attracts the *hau* of birds.[25]

For this entry, I will focus on *mauri* as a key expression of life force. *Mauri* is described in the 'Te Aka Māori Dictionary' as 'life principle, life force, vital essence, special nature, a material symbol of a life principle, source of emotions – the essential quality and vitality of a being or entity. Also used for a physical object, individual, ecosystem or social group in which this essence is located.'[26] In the Māori world, *mauri* is the life force passed down through the genealogies of the *atua* (personified elements/ancestors) to provide life to all known phenomena. It binds the physical and metaphysical together. People, things, a *waiata* (a song), all have their own *mauri. Mauri* is regarded as an attribute of entities to be maintained, nurtured and protected, enabling well-being.[27] *Mauri* can be created by the relationship between things and become a new 'thing'. A grudge between people can have its own *mauri.* Tūhoe *tōhunga* (knowledgeable expert) Hōhepa Kereopa describes how a boil up, a type of Māori stew adapted from colonial cooking methods, has its own *mauri* created by what he describes as the different frequencies of the carrots, onions and meat.[28] We might consider his choice of the word frequency in terms of the energetic character or expression of a 'thing' in this context as an aspect of *mauri.*

Concepts such as *mauri* name the energetic interconnectivity of the human
and non-human worlds. As Hoskins and Jones explain, in the Māori
world it is the relationship between things that is the most important. It
is the energetic exchange rather than the 'thing' itself, that is ontologically
privileged.[29]

Orality

As all 'things' possess *mauri*, a life force that makes them an entity in their
own right, they are thought of as possessing a 'voice' rather than being
emitters of sound. In an oral culture, where knowledge and histories were
encoded in oral forms and shared through generations, where the art of
speaking, singing, poetry and using the 'right' combination of words are
among the highest forms of culture, that non-human 'things' have 'voice'
too demonstrates an awareness of the *mana* (presence and authority) of
the non-human. The 'voice' of 'things', the *mauri* immanent in the sonic
expression of both the human and non-human, is a vehicle for energetic
exchange. As established, *kōauau* are understood to have their own 'voice', to
channel the voices of the *atua* to commune with the spiritual realm, as well
as acting as an extension of the human voice. Nunns explains, 'for Māori
the playing of instruments was intimately connected with song or speech,
and that puoro do not represent an independent "instrument" category of
music but an extension of the human voice – singing, praying, exhorting,
signalling, insulting'.[30] Aesthetic attributes of the human voice can be found
in the performance of the *kōauau* where words such as *hotu* (to sob, pant,
sigh), *tangi* (to cry, mourn or weep) and the vocal equivalent of *wiriwiri*
(to tremble, shiver or shake as a form of vibrato), are used to describe its
sound.[31] Language is introduced to the *kōauau* through the player tongueing
or moving the mouth during performance. Language is a gateway into a
culture and our linguistic abilities impose culturally defined sense and
meaning of the world on our perceptual field. Language is understood
as part of the energetic exchange that maintains well-being – the *mauri
ora* (the well-being of the *mauri*) – between entities in the Māori world.
Jones explains, 'for Māori, language binds speaker and listener (and the
environment) together as they create shared meaning in the moment of their
communication'.[32]

Here I want to focus on the sounding unit of language that forms the core
of a syllable – the vowel. A characteristic of the Māori language is that every

syllable ends in a vowel. Vowels resonate the body in different ways than consonants. The vowel is made by the vocal folds vibrating through the open configuration of the vocal tract with no intervention of the lips or tongue to shape its form. The vocal fold is a part of a muscle on the side of your larynx, also known as the voice box. Like the *kōauau*, it is a hollow, tubular-like passage. The vocal fold connects to the top of the windpipe (trachea) covered with tissues that can vibrate at a high speed, producing sound as air passes through them during exhalation from the lungs.[33]

The onomatopoeia that permeates the Māori language is found in the word *kōauau*. According to scholar and *taonga puoro* player Komene, whose research I have drawn on numerous times in this entry: 'The "*kō*" imitates the sound of a singing *kōauau* voice, and the "*auau*" is the vibrato effect after the note is sounded.'[34] The *aa* to the *uu* is the open space, the sound affect that we have finally come to. Introduced by the initial shape of the consonant, it is that vibrato vowel that holds the sonic space open, it is where the emotion, the intention of the player, the *wairua* (essence, spirit, attitude), can be channelled and where the 'affected', those that hear and respond, can also enter this vibratory portal.

The sonic space held open by the formation of the resonating vowel, in contrast to the phallic-looking *kōauau*, can also be found in the feminine art of the *karanga* (to call). The *karanga* is the ceremonial call performed by women only, that welcomes 'others' into the host's realm by creating a virtual safe passage for them to enter. In the *karanga*, a *kaikaranga* (the women who performs the call) from the hosts and a *kaikaranga* from the visitors, call out to each other in turns. When one call starts to taper off, their phrase performed in one breath, the other begins, and through that overlap of sound, a sonic space is woven and held open through which the guests may safely enter. Hinematau McNeill and Sandy Hata describe the *karanga* as creating a portal between the living and spiritual worlds. The *karanga* strengthens this relational portal, reinforcing the intrinsic link between Ranginui and Papatūānuku, *mana wahine* (the potential/authority of the female) and *mana tāne* (the potential/authority of the male), and the connectivity of the Māori universe.[35] According to NcNeill and Hata, the *whakapapa* for *karanga* lies in the story of the first woman in Māori tradition, Hineahuone. She is recalled in the *karanga* ritual as the first voice on the *marae ātea* (open area in front of the community's communal buildings where the formal welcome to visitors takes place).

Hineahuone begat Hinetītama, representing a life force symbolic of the dawn, often referred to as the Dawn Maiden. When Hinetītama discovered her husband, Tāne, was her father, she fled, horrified, from the world of light to the world of darkness and transformed into Hinenuitepō, the 'Great Woman of the Night'. Hinenuitepō is the *atua* of night and death.[36] Through this narrative, the *karanga* is associated with birth and death, and through *karanga*, the portal between these realms is opened between the material and immaterial, the living and the dead. The *karanga* of an experienced *kaikaranga* is emotionally affecting. Though the words are important, the selection of which are part of the art of the *kaikaranga*, the assertion here is that alongside the relevant words, it is the tone and delivery of the sustained open vowels within which the power to open 'portals' reside. In the context of the *powhiri*, these portals create a safe passage for the guests, while simultaneously opening the thresholds of life and death, to call ancestors to be present among their living descendants. Like the *auau* of the *kōauau*, this entry contends, it is in the sonic space held open by the resonating vowels where the intensity of affect is transmitted.

The tangible and intangible energetics of sound and its affects

At this point we need to establish another 'ground'. What *affect* means is and does takes on different nuances in different contexts. Here I must admit to cherry-picking some concepts and language that have their *whakapapa* in Euro-American posthumanist theories. I also briefly touch on the 'sensing world' of American biologist Lynn Margulis's contribution to Gaian theory to look for language and ideas that might parallel some notion of affect in Western thinking that could be recognized in the Māori world.

In Christoph Cox's Nietzschean/Deleuzean-inspired sonic materialism, when we acknowledge *affect*, we also acknowledge the forces at play in a material world, where the sound event (or any material event) is one where something occurs in the interaction of various kinds of forces.[37] Through Deleuze, we might understand that force equates to any capacity to produce change, whether such capacity and what it produces are 'physical, psychological, mystical, artistic, philosophical, conceptual, social, economic, legal or whatever'.[38] Force always exists in relationship with other forces. All phenomena are expressions and consequences of interactions between forces, with each interaction revealed as an event. In other contexts, we might delve into a more detailed discussion of sonic materialism,[39] but for now, we can use this simple analysis to understand that sound is an aspect of the transformative processes of our material world. We can also use it to establish that sound is part of the transformative processes in the immaterial world where human 'symbolic' life, emotions and energetic traces may be considered extensions of the constantly unfolding material world.[40]

In considering affective relations between entities, I am reminded of the ideas put forward by Margulis regarding the 'sensing world' in her contribution to James Lovelock's Gaian theory. Margulis identifies the way that the interconnected patterns of the systems of the natural world appear to occur in the absence of any central 'head' or 'brain'. Proprioception – a self-awareness of the movement of the body – evolved long before animals evolved and long before their brains did. Sensitivity, awareness and the responses of plants, protoctists, fungi, bacteria and animals, each to their local environment, constitute repeating patterns that ultimately underlies global sensitivity. This sensitivity allows the systems that make up Gaia/earth to self-regulate and maintain conditions that make ongoing life viable.[41] If I can apply the idea of affect here to Margulis's insight, it is understood as a form of sensitivity of the energetic relationships between entities. In the Māori world, sensing ecological systems are part of a network of the extended family that connects through *whakapapa*. As Hoskin and Jones explain, 'the vitality of things is possible not because of the intrinsic qualities of one object alone but because of "its relationship with the mauri [vigour, impetus and potentiality] of others".'[42] Margulis's proprioceptive sensing world folds into an idea of the body's encounter with affect. Returning to a Deleuze and Guattarian understanding, Brian Massumi describes affect as 'the passage from one experiential state of the body to another . . . (with body taken in its broadest possible sense to include "mental" or ideal bodies)'.[43] Something to consider in these modes of thought is the autonomy of the affect. Beyond the interaction of forces, the interfacing event of affect and the affected, something remains. This returns us to the Māori ontological privileging of the relationship between things, how the *mauri* interacts and how that interaction results in *mauri* specific to that interaction. The resonating open vowel sounds of the *kōauau* and a listener receiving them isn't retained in either party but is retained in the resonance of the relationship; that is, in the *mauri* created by this exchange.

In the Māori world there are further terms that resonate with an understanding of affect, namely *ihi*, *wehi* and *wana*. *Ihi* is a psychic power, a form of charisma that elicits a positive psychic and emotional response. *Wehi* is a reaction to that power, and wana composes reactions and aura created during the exchange.[44] These concepts would need their own chapter to draw out their contexts in more depth, but for now, they are described as being named forms of energetical forces that give nuance to affective experiences.

Affect, in the Euro-American formations I've referred to, is generally considered a non-linguistic force. This entry offers – in the context of sound as a 'force' and as an agent of affect – a notion that sound in the Māori world is an extension of the orality and vocality that is understood as being possessed by non-human bodies also. That is, sound and a notion of orality are intertwined. In the context of the *kōauau*, the breath, the *wairua*, the *mauri* of the player and the instrument is carried by Tāwhirimātea – the *atua* of the wind – to the listener, transforming one energy state to another, with bodies acting as conductors and transducers and converting variations in physical quantity to electrical signals to the ear and to the brain. As Komene explains, '*Raukatauri* offers the sound to the world, and *Tāwhirimātea* makes sure her intentions are fulfilled, transmitting the hidden message to the ear of the recipient – human, animal, and insect. The sounds travel to the heart and triggers a feeling or memory that is the same or similar to that intended by the kaiwhakatangi or player.'[45]

In the 'voice' of the *kōauau* – in the *auau*, the *aa* to the *uu* and the *ee ii oo* in between – the open vibratory vowels are a technique to hold open sonic portals leading to powerfully resonant affects. Whether these affects involve entertainment, attracting a lover, procreation, expressing grief, sorrow – calling on *atua* also helps to heal, lessen pain – they exist in the reciprocal interactions of *mauri* between entities, resonating through the past, present and future, to ancestors and back, between the realms of the material and the immaterial.

Tihei mauri ora!

Notes

1 Carl Te Hira Mika, 'The Uncertain Kaupapa of Kaupapa Māori', in *Critical Conversations in Kaupapa Māori*, ed. Te Kawehau Hoskins and Alison Jones (2017; Wellington: Huia Publishers, 2021), 101.
2 Mika, 'The Uncertain Kaupapa of Kaupapa Māori', 100.
3 Mika, 'The Uncertain Kaupapa of Kaupapa Māori', 100. Mika cites T. A. Royal, *Te Ngākau*. Te Whanganui-a-Tara (NZ: Mauriora Ki Te Ao Living Universe).
4 Carl Te Hira Mika, 'The Co-Existence of Self and Thing Through *Ira*: A Māori Phenomenology', *Journal of Aesthetics and Phenomenology* 2, no. 1 (2015): 106.
5 Brian Flintloff, *Taonga Pūoro, Singing Treasures: The Musical Instruments of the Māori* (Nelson: Potter & Burton, 2004), 54.

6 Joël Komene, 'Kōauau auē, e auau tō au e! The Kōauau in Te Ao Māori' (MA diss., University of Waikato, Hamilton, New Zealand, 2009), 57.

7 Brian Flintoff, 'Māori musical instruments – taonga puoro – Melodic instruments – the family of Rangi', Te Ara – the Encyclopedia of New Zealand, http://www.TeAra.govt.nz/en/maori-musical-instruments-taonga -puoro/page-2 (accessed 10 December 2021).

8 Richard Nunns, *Te Ara Puoro: A Journey into the World of Māori Music* (Nelson: Craig Potton Publishing, 2014), 55.

9 auē, e auau tō au e! The Kōauau in Te Ao Māori', 46–7.

10 Leo Fowler, *Te Mana o Turanga: The Story of the Carved House Te Mana o Turanga on the Whakato Marae at Manutuke Gisborne* (Auckland: N.Z. Historic Places Trust, 1974), 24–5.

11 Komene, 'Kōauau auē, e auau tō au e! The Kōauau in Te Ao Māori', 46.

12 Fowler, *Te Mana o Turanga*, 28–9.

13 See Clive Aspin, '*Hōkakatanga – Māori Sexualities*', https://teara.govt.nz/en/ hokakatanga- maori-sexualities/print (accessed 12 December 2021).

14 Peter Beatson, 'Richard Nunns: The Renaissance of Traditional Māori Music', *Music in the Air* 16 (2003): 17–33.

15 Haumanu Collective, '*Haumanu Collective*', https://www.haumanucollective .com/ (accessed 10 December 2021).

16 Nunns, *Te Ara Puoro*, 63.

17 Nunns, *Te Ara Puoro*, 60.

18 Komene, 'Kōauau auē, e auau tō au e! The Kōauau in Te Ao Māori', 62. Komene quotes Hirini Melbourne, Maitai, P. (Producer), Wooster, D. (Producer), and Parata, M. (Director), 'Puoro' [television episode], *Waka Huia* (televised 24 July 1994), Auckland: Television New Zealand Limited.

19 Komene, 'Kōauau auē, e auau tō au e! The Kōauau in Te Ao Māori', 62.

20 Mere Roberts and Peter R. Wills, 'Understanding Māori Epistemology: A Scientific Perspective', in *Tribal Epistemologies: Essays in the Philosophy of Anthropology*, ed. Helmut Wautischer (Aldershot and Brookfield: Ashgate, 1998), 45.

21 Manuka Henare, '*Tapu, Mana, Mauri, Hai, Wairua*: A Māori Philosophy of Vitalism and Cosmos', in *Indigenous Traditions and Ecology: The Interbeing of Cosmology and Community*, ed. J. Grimm (Cambridge, MA: Harvard University Press, 2001), 204.

22 Te Kawehau Hoskins and Alison Jones, 'Non-Human Others and Kaupapa Māori Research', in *Critical Conversations in Kaupapa Māori*, ed. Te Kawehau Hoskins and Alison Jones (2017; Wellington: Huia Publishers, 2021), 52. Hoskins and Jones cite Mason Durie, *Mauri Ora: The Dynamics of Māori Health* (Auckland: Oxford University Press, 2001), x.

23 Hoskins and Jones, 'Non-Human Others and Kaupapa Māori Research', 56.

24 Hirini Moko Mead, *Tikanga Māori: Living by Māori Values*, Revised edn (Wellington: Huia Publishing, 2016), 54.

25 Henare, '*Tapu, Mana, Mauri, Hai, Wairua*', 211.

26   John Moorfield, *Te Aka Māori Dictionary*, https://maoridictionary.co.nz/ (accessed 30 October 2021).
27   Mead, *Tikanga Māori*, 53.
28   Paul Moon, *A Tohunga's Natural World; Plants, Gardening and Food* (Auckland: David Ling Publishing, 2005), 109–10.
29   Hoskins and Jones, 'Non-Human Others and Kaupapa Māori Research'.
30   Nunns, *Te Ara Puoro*, 55.
31   Nunns, *Te Ara Puoro*, 45.
32   Alison Jones, *This Pākehā Life: An Unsettled Memoir* (Wellington: Bridget William Books, 2020), 167.
33   'Larynx', *Encyclopædia Britannica*, https://www.britannica.com/science/larynx (accessed 11 December 2021).
34   Komene, 'Kōauau auē, e auau tō au e! The Kōauau in Te Ao Māori', 4.
35   Hinematau McNeill and Sandy Hata, 'Karanga', in *Kia Rōnaki: The Māori Performing Arts* (Auckland: Pearson, 2013), 60.
36   McNeill and Hata, 'Karanga', 60.
37   Christoph Cox, *Sound, Art and Metaphysics* (Chicago: University of Chicago Press, 2018), 27 *passim*.
38   Cliff Stagoll, 'Force', in *The Deleuze Dictionary*, ed. Adrian Parr, rev. edn (Edinburgh: Edinburgh University Press, 2010), 111.
39   See Marie Thompson, 'Whiteness and the Ontological Turn in Sound Studies', *Parallax* 23, no. 3 (2017): 266–82 and Christoph Cox, 'Sonic Realism and Auditory Culture: A Reply to Marie Thompson and Annie Goh', *Parallax* 24, no. 2 (2018): 234–42.
40   Rachel Shearer, 'Te Oro o te Ao: The Resounding of the World' (PhD diss., Auckland University of Technology, Auckland, New Zealand, 2018), 22–8.
41   Bruce Clarke, *Gaian Systems, Lynn Margulis, Neocybernetics and the End of the Anthropocene* (Minneapolis and London: University of Minnesota Press, 2020), 11–12.
42   Hoskins and Jones, 'Non-Human Others and Kaupapa Māori Research', 53. Hoskins and Jones cite Durie, *Mauri Ora*, x.
43   Brian Massumi, 'Notes of the Translation and Acknowledgement', in *A Thousand Plateaus: Capitalism and Schizophrenia*, ed. Gilles Deleuze and Felix Guattari, trans. Brian Massumi (1980; London and New York: Continuum, 2004), xvi.
44   Tāmati Kruger, 'The Qualities of Ihi, Wehi and Wana', in *Nga Tikanga Tuku Iho a te Māori, Customary Concepts of the Māori: A Source Book for Māori Studies Students*, ed. Hirini Moko Mead (Wellington: Victoria University of Wellington, 1984), 228–45.
45   'Kōauau auē, e auau tō au e! The Kōauau in Te Ao Māori', 57.

Entry 5

Sympathetic response

Charlotte Eubanks

Sympathetic response: resonance, understood otherwise. I stand with my three-year-old before the membrane of a large timpani drum. Across the surface of stretched skin, a scattering of fine grains, a sweep of sand or salt. My child wants to touch it. We learn to touch it with our voices. We begin quietly, a low crooning sound. Nothing. I take in more air, vocalize more loudly and begin to raise the pitch. Suddenly, the grains jump into pattern. A group of boxes. Next, holding a higher note: a set of circles. Then waves. With a few minutes of practice, we find the necessary notes, the appropriate volume. Together, we sing octagonal snowflakes of sand-sound. Our voices joined in sonic patterns make the fine grains vibrate ecstatically in a dance of voice-shape. There are widening rings of pattern as other museum goers circle, gather, watch, disperse and gather again. Later, reading the artist's statement, I learn that she and we have been playing with Chladni plates.[1] But, as soon as the first grains jumped into vibrating squares, I had recognized them by another name: these were *kannō* (sympathetic responses).

Over the course of this entry, I use the Buddhist term 'sympathetic response' as a case study for limning the possibilities of sound understood otherwise. Buddhist literature and ritual practice pertaining to the concept of 'sympathetic response' (感応 Japanese: *kannō*; Chinese: *ganying*) – a term we might translate as 'resonance' – provide a rich conceptual reservoir for the theorization of sound, language, body and affect.[2] The concept, as elaborated in East Asian Buddhist literature, represents a complex amalgam of Indian and Chinese thought in which Indic notions of cause and effect (Sanskrit: *karman*) become enmeshed with pre-existing Chinese cosmological understandings of stimulus and response, which are often illustrated in musical terms: the plucking of a string on one zither causes the same string on a nearby zither to resonate. Broken into its two component parts, *kan* refers to the 'stimulus' or 'emotion' that a practitioner awakens in buddhas and *ō* to the 'response' that this feeling generates in those beings.[3] The practitioner, typically through a vocal act, elicits a favourable attitude in the divine being, who sends a sign to confirm that the practitioner's plea has been registered. This general sense of *kannō* as a mutually cohering relationship between practitioner and buddha is shared by most schools of Buddhism.

To unpack the Buddhist sound theory of 'sympathetic response', this entry explores the *Gyosan taigaishū* 魚山蟇芥集, a ritual handbook that has been used as the basic text for sutra chanting in the Shingon school of Japanese Buddhism since at least the twelfth century. Its teachings regarding sound and the soundedness of the cosmos, however, stretch much further back in time. Thus, I contextualize the *Gyosan* teachings through references to practices of Central Asian sutra translation from the first centuries of the Common Era, a close reading of a conceptual treatise on sound by the eighth-century monk Kūkai, and an examination of several passages from the twelfth-century compilation *Songs to Make the Dust Dance*.

I take guidance from Elizabeth Markham, who has remarked that 'indigenous written notation[s]' need to be understood 'as bearer[s] of indigenous musical conceptualization'.[4] Here, we must understand 'indigenous' not in a linguistic sense (the Japanese language) but in a conceptual one (Buddhist sound theory), for in Japan the sutras are composed in a ritual language that, without training, seems unintelligible, what Brandon LaBelle might call 'gibberish', the 'foundational static' out of which worlding arises.[5] Premodern sound theorists in Japan, however, characterized Buddhist chanting neither as gibberish nor as static. Rather, they speak of pattern (文字, Jp: *monji*), of vibration (響き, Jp: *hibiki*) and of dust (塵, Jp: *chiri*), and their interactions coalesce into miraculous, but utterly natural, sympathetic responses. These are the terms I privilege in the following pages, for they suggest a way of understanding anew the power of sacred sounds to make what is always present, immediately apparent.

Pattern/文字/*monji*

The monk Kūkai (774–835 CE) produced one of the most extensive bodies of Buddhist scholarship on sound. During his brief but productive stay at the intercultural entrepôt of Chang'an, Kūkai studied with a wide range of individuals from the Indian subcontinent, Inner Asia and Tang China, reading texts in both Chinese and Sanskrit. After returning to Japan, Kūkai synthesized his learning in several essays, dwelling at length upon the affective capacities of sound. Perhaps the most central of Kūkai's writings, sometimes referred to as his 'trilogy' (*sanbu no sho*),[6] are the three treatises *Attaining Enlightenment in This Very Existence* (Sokushin jōbutsu gi), *The Meanings of*

Sound, Word, and Reality (Shōji jissō gi) and *The Meanings of the Word Hūṃ* (Unji gi), which have been mandatory reading for Shingon Buddhist monks for the last millennium.

Kūkai's primary concern in these essays lies in what he calls *monji*. He opens *On the Meanings of Sound, Word, and Reality* with a crucial assertion: 'The Enlightened Being explains the dharma by means of *monji*. And these *monji* are constituted by the six kinds of objects' (夫如來説法必藉文字。文字所在六塵其體).[7] While in modern usage the term *monji* might typically be translated as 'word' and, more particularly 'written word', Kūkai is not arguing here that enlightened beings depend upon writing to convey their teachings. Instead, he asserts that *monji* must be understood more broadly as sense objects: all that can be seen, heard, smelled, touched, tasted or thought. Thus, we can gloss the word *monji* as a 'pattern' that human beings can perceive sensually.

Continuing, Kūkai notes that the enlightened being shows the deluded how to return to the path of correct perception and 'to do so must use sounds and words (*shōji*). When sound and word are clear, then true reality (*jissō*) will be revealed' (名教之興非聲字不成。聲字分明而實相顯所謂聲字實相者).[8] While Buddhist sacred texts (sutras) most commonly describe enlightened beings communicating via spoken words (sermons), buddhas do regularly employ other methods of patterning, ranging from smell (producing the scent of incense or flowers) to sight (emitting rays of light from the forehead), touch (placing the hand on the earth) or taste (presenting listeners with fragrant rice to eat). While acknowledging these other sense patterns, Kūkai is most concerned with sound. 'No sooner does the breath issue forth [from the mouth] than it touches the air, inevitably producing the vibration that we call "sound" (*shō*). Sound names a thing and this we call "word" (*ji*). The word summons up the thing named and this we call "reality" (*jissō*)' (內外風氣纔發必響名曰聲也。響必由聲。聲則響之本也。聲發不虛。必表物名號曰字也。名必招體。名之實相).[9] Because, by definition, enlightened beings' communications are true, their utterances do not only describe reality; they summon it into being, making what is (always, already) present apparent. Kūkai declares that the 'five notes and the eight notes' of the Chinese musical scales and the 'seven declensions and eight cases' of classical Sanskrit grammar constitute the forms of sound that are most efficacious, in that they have the strongest summoning ability (五音・八音・七例・八轉。皆悉待聲起).[10] As Kūkai argues later in the treatise, the 'five great elements' – earth, air, fire, water and emptiness: all that provide the material substrate of lived reality – 'all have vibrations' (五大皆有響),[11] and these vibrations make patterns. 'Vibration' here suggests not just what is heard but what is felt: a sound that echoes and reverberates, to be sure, but also one that resonates within the sense body, shaking it, vibrating it, attuning it.

Whereas all of reality literally hums, the matter of patterning (*mon*) is crucial, and the presence of patterning is what distinguishes sound from noise. 'All the languages in all the worlds arise from sound (*shō*). In sounds there are long and short [vowels], high and low [tones], and various inflections which, taken together, comprise a pattern (*mon*)' (此十界所 有言語皆由聲起。聲有長短高下音韻屈曲。此名文。文由名字。名字待文).[12] In other words, the vibrational pattern of the voice, when enunciated accurately, comprises a pattern that has the power to stir what it names. Not all sounds, however, achieve this. Deluded beings produce 'faulty' (妄, Jp. *mō*) sounds, while enlightened beings produce sounds that are 'accurate' or 'real' (眞實, Jp. *shinjitsu*).[13] Correctly patterned sounds, also known as mantra (Jp: *mandara* 曼荼羅),[14] possess vibrational power. So central was this claim to Kūkai's teachings that his school, Shingon, takes its name from the 'true speech' (Jp: *shingon* 眞言) that he describes here as being 'real'.[15] Not surprisingly, given his intensive theorization of patterning, Kūkai summarizes his teachings in this treatise in the form of a sonically patterned, five-character quatrain (Ch: *jueju*). 'The five great elements: each has vibration/The ten realms: all have language/The six sense fields: each has expressive pattern/The absolute buddha body: this is true reality' (五大皆有響 十界具言語 六塵悉文字 法身是實相).[16] What Kūkai is doing in this poem, and in his Sinitic writings in general, is to leverage the 'patterned harmony' of Chinese poetry to transduce the potency of Sanskrit sound into a different linguistic setting with radically different syntax.[17] The key to this transduction is what Buddhist sound theorists call 'vibration'.

Vibration/響き/*hibiki*

In East Asia, reading Buddhist texts is neither a simple nor a straightforward proposition. In the early centuries of the Common Era, Central Asian monks rendered Buddhist scriptures into a host of languages, including Chinese. These monks chose to translate some terms, while deciding to transliterate others.[18] The first Buddhist texts reached Japan in the form of these classical Chinese translations, and so they have remained, despite an awareness of both the presence of Sanskrit originals and the capacities to translate them into local vernaculars. Furthermore, in moving from a tonal language (Chinese) to an atonal one (Japanese), Buddhist reciters needed to accommodate vocal inflections, eventually coming to understand Sinitic tones governing semantic meaning as musical notes and glissandos.[19] To recite Buddhist texts in medieval Japan thus required one to vocalize phonologically Japan-icized transliterations of a Chinese translation already studded with Sinitic transliterations of Sanskritic loan words. The upshot of this complex situation is that there are strong spiritual and secular motives mandating the creation of a recognized class of skilled reciters (Jp: *nōdoku* 能読), that is, ritual specialists whose area of expertise lies in mastering, performing and transmitting to the next generation acoustically accurate, ritually effective vocalizations of sacred text, a practice known as 'illumination of the voice' (Jp: *shōmyō* 聲明).

Mount Kōya, the monastic centre established by Kūkai in the ninth century, was and remains one of the major centres of *shōmyō* practice in Japan. Its teachings are based on Kūkai's concepts of the relation between voice, sound and 'pattern' discussed above and are synthesized in the ritual handbook known as the *Gyosan taigaishū*. A comprehensive three-volume imprint of the *Gyosan* was first published in 1646,[20] but its contents are based on much older, orally transmitted instruction and draw heavily on the teachings of a monk named Kakui (1237–93 CE).[21] Kakui, Kūkai and indeed everyone associated with *shōmyō* recitation in Japan understood that the sacred texts they were reciting were comprised of Chinese translations of South Asian texts, and as such, the recitation texts do not replicate Sanskrit sounds. What they do claim to replicate is the vibrational efficacy of those Sanskrit sounds.

An eighteenth-century preface to the *Gyosan*, authored by Shingen Kinshiki (fl. ca. 1729 CE), himself a monk and *shōmyō* practitioner based on Mount Kōya, tackles the matter of translingual transduction head-on. He writes, 'As for the *Gyosan taigai*, though there are distinctions to be made between esoteric and exoteric teachings, both extol the auspicious vibration (瑞響, *suikyō*) of Chinese. In exoteric teachings, you have translations of what is called the golden speech (金言, *kongen*)' – that is, the teachings of the historical Buddha as he spoke them. 'But', he continues, 'there is some question as to whether or not they carry across the Sanskrit vibrations (梵響, *bonkyō*)'.[22] He asserts that exoterically transmitted Chinese musical arrangements enable the translated texts to be chanted in such a way as to retain the vibrational qualities of the Sanskrit. 'Moreover', when done correctly, 'the myriad mysteries of Sanskrit vibration manifest in the brilliant radiance of the numinous Japanese pronunciation' (訓の霊變粲然)[23] of the Chinese written characters. Finally, he concludes, we know that *shōmyō* recitation effectively transduces the vibrational power of the Buddha's teachings because, 'through these vocalizations, all living things achieve enlightenment' (衆生ハ又音聲ニ由テ領悟ス).[24]

Immediately following Shingen Kinshiki's preface, the *Gyosan* manual includes a set of diagrams suggesting how this system of musical vibration works. While I will not delve deeply into the musicology of the matter here, a quick look at these diagrams does provide some sense of how all-encompassing this sonic system is. The first diagram is what musicologists call a 'circle of notes', here constellated around the character *ji* (字), a generic placeholder for any written word, around which musical annotations might accrete (see Figure 5.1). Of course, the use of *ji* (字) also suggests Kūkai's notion of *monji* (文字), not simply as 'writing' but, more broadly, as any pattern that human beings can perceive sensually. The circle of notes contains three groupings of five notes each. The *Gyosan* suggests that musically trained humans can produce eleven of these notes. The four unproduceable notes, along the right-hand side of the image, are indicated by being blank, and the three groupings are indicated by the larger brackets at the top-right, bottom-right and left-hand side.[25] While some of the notes marked in *shōmyō* treatises as humanly irreproducible may have been achieved partially, by means of vocal overtones or through the use of musical instruments (bells and drums of various sorts), it is important to recognize here that the human voice is fundamentally embedded in, and part of, a larger cosmos, such that the entire sonic atmosphere sounds together. The placement of *shōmyō* practice halls in heavily wooded areas, typically adjacent to small waterfalls, means that the human voice blends with the susurration of wind through leaves, the high notes of birdsong and the roar of falling water.

Further, the circular arrangement of the standard pitch diagrams is no accident. The human voice is not understood to move up or down a scale envisioned as ascending ever higher on the one extreme and descending ever lower on the other. Instead, the human voice is part of repeating cosmic cycles. These markings can be used, by the appropriately trained musician, as templates for generating a set of 'fundamental chromatic notes in a tone-system' that links to a broader set of ritual, metaphorical, environmental, medical, locational and seasonal corollaries.[26] A second diagram in the *Gyosan* names some of these interconnected nodes (see Figure 5.2). The note *C* (宮) correlates, among other things, to 'centre' (中央) and the 'Great Sun' buddha Dainichi (大日, Vairocana). The note *D* (商) connects to the cardinal direction west (西), the season of autumn (秋), the buddha Amida (彌陀, Amitābha) and so forth, until we come to the note *A* (羽), which connects to north (北), winter (冬) and the historical Buddha (釈迦, Śākyamuni). Ritual specialists using the *Gyosan* thus understand sound as interrelated with the movement of heavenly bodies, the turning of the seasons, the health and decline of the human body, the ritual calendar of the court and of religious centres, the salvific positionality of one's current self, the aggregate building blocks of material reality and so forth. Acting on one of these nodes – producing a certain vocal tone, for instance – was understood to

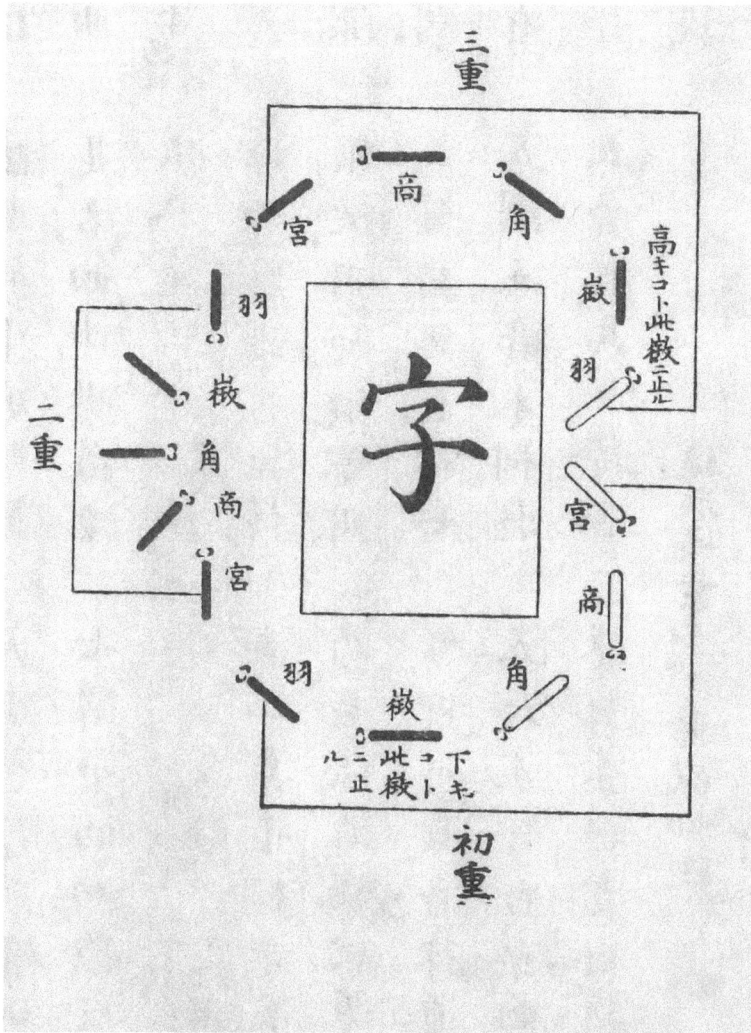

Figure 5.1 The circle of notes from the opening pages of *Gyosan taigaishū* (Ashihara Jakushō, ed.: Wakayama-ken: Kimura Tomematsu, 1892 imprint).

act simultaneously on each of the other nodes in a way that was verifiable (or was, at the very least, believed to be verifiable) upon phenomenological introspection or empirical observation of the external world.[27] Extending Five Phase (Ch: *wuxing*, 五行) cosmological theory, Buddhist musical

○
本調子事

宮　中央土用　大日　一越　呂四
高　西秋　彌陀　平調　律五
角　東春　阿閦　雙調　呂一
嶽　南夏　寶生　黃鐘　中曲半呂二　半律二
羽　北冬　釋迦　盤渉　律三

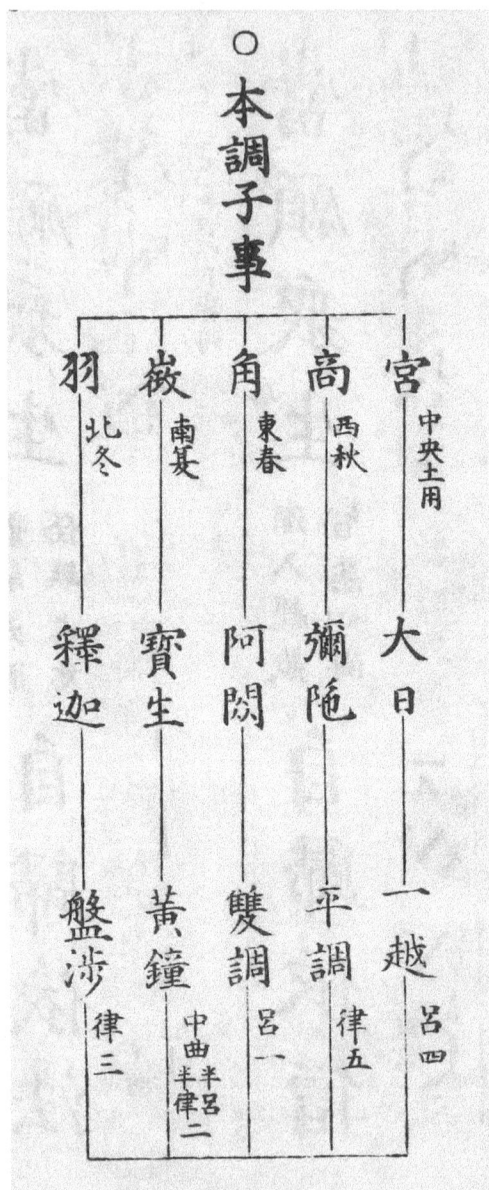

Figure 5.2 Chart showing correlations between notes, seasons, directions and deities in *Gyosan taigaishū* (Ashihara Jakushō, ed.: Wakayama-ken: Kimura Tomematsu, 1892 imprint).

treatises and ritual handbooks assume the human voice is directly keyed to the movement of the cosmos. And the cosmos itself is made of 'dust', dust which can be made to dance.

Dust/塵/*chiri*

One of the most useful texts for understanding how exoteric Buddhist notions of the soundedness of the cosmos (such as those expressed in the *Gyosan*) permeated medieval Japanese culture is the *Ryōjin hishō* (梁塵秘抄, songs to make the dust dance on the beams). A collection of lyrics compiled by Emperor Goshirakawa (1127–92 CE) in the twelfth century, the anthology consists of two parts: *Kashishū* (歌詞集, the lyrics) and *Kudenshū* (口伝集, notes on oral instruction). While the extant text is missing important prefatory material and colophons that might allow us to contextualize it more precisely, the end of volume one contains a brief passage explaining the title of the work. Like many works from classical and medieval Japan, the text borrows some prestige from the continent. Here, rather than referring to the words or actions of an emperor, bureaucrat or general, however, the collection points to two women renowned for their singing: Yu Kung and Han E. The *Ryōjin hishō* states that their 'voices were wonderful and mysterious, well beyond compare', moving all who heard them to tears. Indeed, the text continues, 'The resonance of their singing voices stirred up the dust on the roofbeams and it did not settle again for three days, and so this collection has been named the secret writings (*hishō*) of the dust (*jin*) from the roofbeams (*ryō*)'.[28]

While 'dust' (*jin*, *chiri*) can certainly refer to a fine powder – those tiny particles of matter that gather on surfaces or are lofted on currents in the air – the word carries more particular connotations in East Asia, where Buddhist translators often employed this word to translate the Sanskrit terms *artha*, *viṣaya* and *gocara* into Chinese. The Sanskrit terms bear important religious, philosophical and scientific import, referring to the material objects that make up the world of lived reality. More precisely, *viṣaya* connotes objects of cognition, all that is perceived by the sense organs (eye, ear, nose, mouth, touch and mind). *Artha* and *gocara* additionally connote a certain uncleanliness because sensual perception typically is regarded as defiling the mind by giving rise to desire.[29] Given the semantic range and the Buddhist connotations of the word 'dust', the title *Ryōjin hishō* might aptly be translated as 'A treasured selection of songs that shape reality', for that is what the singers Yu Kung and Han E were doing.

Tellingly, the *Ryōjin hishō* brackets the explanation of its title with a set of lyrics that are variations on the theme of enlightenment. Immediately preceding the passage, we have the verse, 'Though it may seem that Śākyamuni/attained perfect enlightenment in this incarnation/you, too, have been a buddha/for five hundred dust world kalpas' (*Shaka no shōkaku naru koto wa/kono tabi hajime to omohishini/gohyaku jitengō yorimo/anata ni hotoke ni naritamafu*).[30] Following the passage explaining the collection's title, we have the lyric, 'Though it may seem that Śākyamuni/attained perfect enlightenment in this incarnation/you, too, have been regarded as a buddha/for five hundred dust world kalpas' (*Shaka no shōkaku naru koto wa/kono tabi hajime to omohishini/gohyaku jitengō yorimo/anata ni hotoke to mietamafu*).[31] The 'five hundred dust world kalpa' phrase is taken from the sixteenth chapter of the *Lotus Sutra*.[32] This chapter explains that Śākyamuni Buddha originally attained enlightenment not in his most recent incarnation but, in fact, innumerable aeons in the past. The 'dust' of the phrase 'dust world kalpas' is part of an elaborate metaphor that emphasizes the enormity of time during which Śākyamuni has been preaching to sentient beings. The 'dust' also suggests the material world of old age, sickness and death in which he has been labouring. More crucially, though, the two songs perform (or announce) an almost magical act of time-bending: what you thought lay in the recent past in fact occurred aeons before; what you thought lay in the distant future is (*naritamafu*) and is already apparent (*mietamafu*) now.

Sympathetic response/感応 /*kannō*

These verses on 'dust' return us, then, to the notion of 'sympathetic response' insofar as they collapse time, suggesting that the singing body, right here and right now, *is* the body of perfect enlightenment. To stir up 'dust', therefore, means to polish or clear off what other verses call the 'grime of life and death' (*seishi no chiri*)[33] and to draw attention to an object of cognition or perception, thereby exciting the senses into an enlightened state. Reading the lyrics together with the inscription they bracket, we understand that Yu Kun and Han E's 'chanting voices' (*utahikeru kowe*) are pitched perfectly to 'resonate with' (*hibiki*) the matter of the universe (*chiri*) and to alter it, shifting its location and causing it to linger in the air, held aloft in a time-bending moment of enlightened existence.

Indeed, Emperor Goshirakawa, compiler of the *Ryōjin hishō*, believed that singing – his singing – had the power to manifest the divine. Shortly after the turning of the New Year in the first month of 1169 CE, the emperor went on pilgrimage to Kumano Shrine. He was anticipating his formal retirement from politics and assumption of the tonsure. Deciding to spend the night alone in the main shrine hall, he had his men bed down nearby, where he could still hear them or call out to them as needed. Accompanied only by candlelight, he chanted sutras and sang throughout the night until about 4.00 am, when his voice finally achieved the resonance for which he had been aiming. Describing the scene in the pages of the *Ryōjin hishō*, he writes: 'As dawn approached, everyone in the group grew still and, when not a human sound was to be heard, I settled my heart and sang . . . No sooner had I done this than there came an incomparably musky scent wafting in from the direction of the two main shrines.' The scent is that of the musk deer (*Moschus moschiferus*), whose glands provided an ingredient in expensive perfumes and in offerings of incense. Two of the emperor's noble attendants comment on the auspicious scent and, 'while all gathered there were still wondering over this, there was a rumble as if the building itself were shaking'.[34] When his companions express concern, the emperor observes that the sound may be coming from 'birds who have taken up roost for the night in the rafters' of the building. The suggestion, of course, is that his own singing has not only agitated the dust that has come to rest on the roofbeams but, in so doing, has also wakened the birds from their slumbers.

The rustling of the birds sets off a chain reaction. 'Immediately, the [musky aroma] permeated the area and the blinds in the sanctuary lifted, as if someone were entering the room. Even the sacred hanging mirror rocked side to side and, for quite some time thereafter, everything shook.'[35] Goshirakawa suggests here that his singing has stirred up the cosmos, causing the deities of the shrines (Buddhist avatars) to manifest themselves, their presence confirmed by smell (the musky aroma), touch (the feel of the breeze, the shaking of the floor), sound (the rustling in the eaves) and sight (the rocking of the mirror and lifting of the blinds).

Though he does not specify which song he was performing at the time, any number of lyrics included in his collection promise just this type of divine resonance. A typical verse, which essentially reprises the above scene, runs thus: 'When, in the silence of the temple hall/one offers flowers and incense to the buddhas/quieting one's heart and reciting the scriptures for a time/there! Buddhas appear' (*shizukani oto senu dōjo ni/hotoke ni hana kō tatematsuri/kokoro shizumete shibarakumo/yomeba zo hotoke wa mietamafu*).[36] The verse asserts that, through singing, one can render the buddhas sensually perceptible. The word *shizumu* (静む) means 'to be quiet' and is a homophone with *shizumu* (沈む), meaning 'to sink'. The idea is that – just like dirt settles in still water, revealing the water's clear, translucent nature – when the agitations of the heart 'sink', the buddha (within and without) is naturally revealed. Some 'dust' needs sound to dance, while other 'dust' needs silence to settle, so what is present may become affectively apparent.

Notes

1 The plates are named for Ernst Florens Friedrich Chladni (1756–1827 CE), the physicist and musician who created them for his research on acoustics. For artist Meara O'Reilly's statement, see the permanent exhibit's page at https://mearaoreilly.com/Chladni-Singing (accessed 26 February 2021).

2 For other English-language work on sympathetic response, see Robert H. Sharf, *Coming to Terms with Chinese Buddhism: A Reading of the Treasure Store Treatise* (Honolulu: University of Hawai'i Press, 2002), 77–133. See also Charlotte Eubanks, 'Sympathetic Response: Verbal Arts and the Erotics of Persuasion in the Buddhist Literature of Medieval Japan', *Harvard Journal of Asiatic Studies* 72, no. 1 (2012): 43–70.

3 I refer to enlightening beings generally as 'buddhas', while the capitalized 'Buddha' refers to the historical Buddha Śākyamuni.

4 Elizabeth Markham, translators' notes, in Leopold Müller, *What the Doctor Overheard: Dr. Leopold Müller's Account of Music in Early Meiji Japan (Einege Notizen über die japanische Musik, 1874–1876)*, ed. and trans. Elizabeth Markham, Naoko Terauchi, and Rembrandt Wolpert (Ithaca: Cornell East Asia Series, 2017), 48.

5 Brandon LaBelle, *Lexicon of the Mouth: Poetics and Politics of Voice and the Oral Imaginary* (New York: Bloomsbury, 2014), 67.

6 Yoshito S. Hakeda, *Kūkai: Major Works* (New York: Columbia University Press, 1972), 77. For Hakeda's translations of the three treatises, see 225–62. For an analysis, see Ryūichi Abé, *The Weaving of Mantra: Kūkai and the Construction of Esoteric Buddhist Discourse* (New York: Columbia University Press, 1999), 275–304.

7 For the full text, see Junjirō Takakusu, Watanabe Kaigyoku, and Ono Gemmyō, eds., *Taishō shinshū daizōkyō* (Tokyo: Taishō Issaikyō Kankōkai, 1924–34), 77: text number 2429. This quotation is from page 401, register c, lines 7–8. Hereafter, T2429.77 with the page and line numbers following.

8 T2429.77.0401c12–13.

9 T2429.77.0401c20–22.

10 T2429.77.0401c24–25.

11 T2429.77.0402b10.

12 T2429.77.0402b25–27.

13 T2429.77.0402c01; T2429.77.0402c02.

14 T2429.77.0402c03.

15 T2429.77.0402c05.

16 T2429.77.0402b10–11.

17 On 'patterned harmony', see Jonathan Stalling, *Poetics of Emptiness: Transformations of Asian Thought in American Poetry* (New York: Fordham University Press, 2010), 61–95. On transduction, see Patrick Eisenlohr, *Sounding Islam: Voice, Media, and Sonic Atmospheres in an Indian Ocean World* (Berkeley: University of California Press, 2018), 79–108.

18 For a survey of some ways in which Chinese deals with translation, tran-
    scription and loan words, see Ping Chen, 'Modern Written Chinese in
    Development', *Language in Society* 22, no. 4 (1993): 505–37. Though Chen
    deals with the contemporary era, many of the techniques were developed in
    the pre-modern period.

19 For more on the musical affordances of Buddhist texts, see Charlotte
    Eubanks, 'Cultures of Sound: Lineages and Languages of Sutra Recitation
    in Goshirakawa's Japan', in *The Languages of Religion*, ed. Sipra Mukherjee
    (London: Routledge, 2018), 17–34.

20 Jakushō Ashihara, ed., *Gyosan taigaishū* (Wakayama-ken: Kimura Tome-
    matsu, 1892). Hereafter, *Gyosan*. This edition is a revised reprint from
    the Meiji period (1868–1912). In all instances, I have checked this reprint
    against earlier versions of the text, including the 1685 version owned by the
    National Institute of Japanese Literature.

21 Kōken Arai, '*Shōjitsushō* ni tsuite: *Gyosan taigaishū* to no hikaku o chūshin
    ni', *Bukkyō bunka gakkai kiyo* 24 (November 2015): 103–40. For further
    information, including a historical overview of *Shingon shōmyō kuden*, see
    Atsuko Sawada, 'Shingon shōmyō no kudensho ni tsuite 2: *Shōjitsushō*',
    Osaka kyoiku daigaku kiyo 35, no. 2 (December 1986): 231–43.

22 *Gyosan*, vi a–vi b. In this edition, each spread (rather than each page) is
    numbered. Thus, vi a refers to the right-hand page of the sixth spread of the
    preface.

23 *Gyosan*, vii a.

24 *Gyosan*, vii a.

25 This basic musical conception is shared across schools of *shōmyō*, and it
    is common to many classical musical forms in the Japanese archipelago.
    On classical musical notation, see Ryūsho Wagatsuma, 'Sandaihakkenki ni
    kansuru ikkōsatsu: Sōchō hon'i goon sanjū kenritsu monsetsu o chūshin ni',
    Chisan kanga Kūkai 64 (2015): 55–71.

26 Markham, Terauchi, and Wolpert, translators' notes, in Müller, *What the
    Doctor Overheard*, 21.

27 For more on the proximity of music to cosmology, see Kenneth J. DeWo-
    skin, *A Song for One or Two: Music and the Concept of Art in Early China*
    (Ann Arbor: University of Michigan Press, 1982). See also Erica Brindley,
    Music, Cosmology, and the Politics of Harmony in Early China (Albany: State
    University of New York Press, 2012). On early Sinitic conceptions of the
    voice, see Judith T. Zeitlin, 'From the Natural to the Instrumental: Chinese
    Theories of the Sounding Voice before the Modern Era', in *The Voice as
    Something More: Essays towards Materiality*, ed. Martha Feldman and Judith
    T. Zeitlin (Chicago: University of Chicago Press, 2019), 54–76. See also
    Kenneth J. DeWoskin, 'Philosophers on Music in Early China', *The World of
    Music* 27, no. 1 (1985): 33–47.

28 Hisao Kawaguchi and Nobuyoshi Shida, eds., *Wakan rōeishū, Ryōjin hishō,
    Nihon Koten Bungaku Taikei* (Tokyo: Iwanami Shoten, 1965), 73: 345. Here-

after, NKBT 73. For an English commentary, see Yung-Hee Kim, *Songs to Make the Dust Dance: The Ryojin Hisho of Twelfth-Century Japan* (Berkeley: University of California Press, 1994).

29 See A. Charles Muller, *Digital Dictionary of Buddhism*, http://www.bud-dhism-dict.net/cgi- bin/xpr-ddb.pl?q=塵 (accessed 4 September 2020).

30 NKBT 73: 344.

31 NKBT 73: 346.

32 For a full English translation, see Leon Hurvitz, trans., *Scripture of the Lotus Blossom of the Fine Dharma, Translated from the Chinese of Kumārajīva* (New York: Columbia University Press, 1976).

33 NKBT 73: 368.

34 NKBT 73: 463.

35 NKBT 73: 463.

36 NKBT 73: 361.

Entry 6

Whispers

Christian de Mouilpied Sancto

Whispers carry a distinctive range of affects. Figuratively, whispering is synonymous with gossiping or spreading rumours, while literal whispers can readily induce pleasure – witness the popularity of ASMR – or be harbingers of danger, as with the malevolent whispering spirits of many a horror film. This entry studies how, in his 2017 sound installation *Whispering Campaign*, the artist Pope.L leveraged the affective qualities and political associations of both figurative and literal whispering.[1] Commissioned for the fourteenth edition of documenta, titled 'Learning from Athens', *Whispering Campaign* was installed in various locations at the art quinquennial's two host cities, Kassel and the Greek capital. By the time he received the commission, Pope.L had been exhibiting widely in the United States for roughly four decades. While his work across media including performance, painting, drawing, installation and theatre has attracted much attention in his home country for its idiosyncratic take on race relations, *Whispering Campaign* was Pope.L's first major commission from a European art institution.[2]

The idea of making a work based on whispering came to the artist early. In advance of a research trip to Greece in spring 2016, Pope.L emailed his proposal to documenta staff: 'Imagine a set of whisperers who whisper twenty-four hours a day for one hundred days – what are they whispering? Secrets.'[3] The 'secrets' would eventually materialize as a script comprising short, heterogeneous texts in Greek, German and English collected by the artist and German and Greek assistants. The script was then recorded in whispers by voice actors in Athens, Kassel and Chicago, Pope.L's city of residence. During documenta, the recordings were played back through speakers installed in thirty-one locations across Kassel and Athens, including restaurants, public parks, shopping centres and back alleys, as well as documenta's gallery spaces. Further material was provided by a group of live performers who, in a series of three-hour performances, wandered individually around designated areas of the cities whispering both scripted and improvised texts.

While one curator described *Whispering Campaign* as 'among documenta 14's most talismanic' works, many commentators found the piece to be confusing, disorienting and even threatening. Artist Chichan Kwong relayed how 'Not knowing when, where, and how a whisper would happen made me want to attribute even the tiniest noise to the campaign', while curator Dieter Roelstraete remarked how the whispers made it seem 'as if the entire urban fabric of Kassel had been invaded, infiltrated, contaminated'.[4] Others voiced concern about one short phrase from the long script: 'Ignorance is a virtue.' Some members of the public responded aggressively upon hearing the phrase, according to the live whisperers. The performers also expressed their own anxieties about appearing to ventriloquize the political philosophy of the newly elected US president Donald Trump, especially in the 'socially more diverse and challenging area[s]' of the city.[5] Indeed, curator Monika Szewczyk traced the phrase back to an indirect exchange between Trump and Barack Obama in May 2016, when the former's presidential campaign was in full swing. At Rutgers University's commencement address, Obama ridiculed the notion that disregarding factual knowledge could be a viable political strategy. 'Ignorance is not a virtue', he advised the graduating seniors, a motto that immediately made headlines. The following day, Trump quoted the phrase in a tweeted retort: '"In politics and in life, ignorance is not a virtue." This is a primary reason that President Obama is the worst president in US history!'[6]

Trump had conducted his election campaign on the wager that when it comes to getting votes, facts matter less than emotions, but *Whispering Campaign* makes reference to such election strategies beyond its 'Ignorance is a virtue' slogan. A whispering campaign is a persuasion strategy in which political groups inject mendacious rumours into a populace in order to soil their opponents' reputations. For example, in the 2000 Republican presidential primary, a whispering campaign perpetrated against John McCain in some southern US states insinuated that the senator's adopted Bangladeshi daughter was in fact a Black child McCain had fathered out of wedlock.[7] But whereas the whispering of whispering campaigns is figurative (most of the time, people spreading rumours don't *literally* whisper to each other), Pope.L yokes *actual* whispering to its figurative associations. And as the comments quoted earlier testify, the array of affective responses to the work resulted from diverse interpretations of the recited texts and whispers' sonic characteristics.

The texts comprising *Whispering Campaign*'s script are heterogeneous in style, content and tone. Some of the script's most poignant passages relate to the so-called migrant crisis that was then a salient political issue in both Greece and Germany. Klea Charitou recounts the provenance of one such text: a transcription of the story of one Mondol Rashel, a young man the artist met on a street outside the Athens Polytechnic. Rashel was selling tax-free cigarettes on the street and sending half of the proceeds back to his family in Bangladesh. He said he hoped to be able to work legally in Greece one day and to eventually return to a normal life back home. Pope.L transcribed the migrant's story and juxtaposed it with the tale of Agamemnon, the itinerant mythical Mycenaean king. In so doing, says Charitou, 'the whispers align ancient myths with contemporary trials, the continuity of Agamemnon and Rashel bound in cryptic, dark connections, uncovering truths of sacrifice and migration, of survival and vengeance. . . . The hitherto anonymous migrant Rashel is transformed by Pope.L into a contemporary Homeric hero whose dreams of a better life are cloaked in sadness.'[8] But in the final work, Rashel remains anonymous; his name is not included in the script or anywhere else besides Charitou's recollections, which were published several years after documenta 14. What's more, Rashel's anonymized exploits are irreverently inserted into the work's script amidst descriptions of buildings, pithy slogans, sequences of seemingly random numbers and whimsical appropriations of German and Greek literary history, such as 'Ferdinand Grimm stumbles out of the narrow but welcoming doorway of the Night Time 24 Hour Bar and Grill and promptly vomits all over my pink pants', all of which were whispered with the same measured, affectless delivery.[9]

Lauren Berlant identifies the dialectic of affect and affectlessness in Pope.L's work as a 'deadpan aesthetic'. Borrowing David Robbins's words, she describes deadpan as that which 'display[s] the emotional neutrality of data yet retain[s] an existential charge of theater', thereby interpellating its audience into an 'absurd intimacy'.[10] *Whispering Campaign* transposes this deadpan aesthetic into the realm of sound, its 'absurd intimacy' amplified by the work's installation in various urban spaces (including that institution of awkward public intimacy par excellence – the public restroom). In this case, however, deadpan is manifested not in embodied performance or visual objects, but via a mode of vocal mediation with its own phenomenological particularities: the acousmatic voice.

By separating sounds (the whispers) from their sources (the whisperers), *Whispering Campaign* partakes in what has come to be known as 'acousmatic' sound. The term 'acousmatic' derives from Pythagoras's conceit of delivering lectures from behind a curtain, the theory being that his students would better concentrate on his words if the visual aspects of his performance were removed.[11] The notion regained currency in the age of recorded sound. In the mid-twentieth century, Pierre Schaeffer used acousmaticity to theorize the experience of listening to electronic music such as his own *musique concrète*. Echoing Pythagoras, Schaeffer held that without seeing the sources of sounds, listeners may better attend to their sonic characteristics.[12] More recently, critical theorists have begun to emphasize the affective and political dimensions of acousmatic sound, and of the acousmatic voice in particular. In his 2006 volume *A Voice and Nothing More*, Mladen Dolar describes the acousmatic voice – a voice that is disconnected from a body, for example, by radio, telephone or sound recording technology – as affectively disturbing. 'Because it cannot be located', he writes, the acousmatic voice 'seems to emanate from anywhere, everywhere'; it is 'uncanny', 'haunting', 'spectra[l]'.[13] More recently still, Rey Chow has theorized acousmaticity as a site of sonic and subjective multiplicity. For her, sound reproduction technologies separate sounds from sources and subject them to the aura-vanquishing powers of the mechanical (or digital) copy. Acousmatic sound thus entails a 'proliferating multitude of causes and sources, a multitude that renders the question "where are these sounds coming from?" by and large meaningless'.[14] This 'multitude of causes' in turn heralds 'the evisceration of a viably unified and authoritative point of command, of what some might call a transcendental subject position'.[15] In short, whereas traditionally acousmatic techniques were used to focus attention on a single sound source, more recently they have been used to create the impression of multiple sound sources, the results of which are legible in both affective and political registers.

Although Dolar argues that in today's mediascape the acousmatic voice has become a banal commonplace, *Whispering Campaign*'s listeners suggest that the work reawakened its capacity to elicit disturbance.[16] Adjectives like 'uncanny', 'spectral', and 'haunting' pepper commentaries on the work, while curator Leon Hösl mused that 'the whispers seem to belong to the city', as if confirming Dolar's observation that the acousmatic voice 'seems to emanate from anywhere, everywhere'.[17] Moreover, by deciding to use whispering, an especially quiet kind of speech, in a work installed in urban public spaces, Pope.L made his work particularly reliant on acousmatic mediation. What would ordinarily be a private, intimate act is made audible in urban settings by acousmatic technologies like microphones, speakers, PA systems, and the radio.[18]

But a more figurative level of acousmaticity resides in the anonymity that pervades *Whispering Campaign* – that of the authors of the text fragments, as well as of the whisperers themselves. The script does not attribute the fragments to any sources. In certain cases, we might infer something of the author's identity from the text itself – we might imagine, for example, that the account of fleeing Syria for a refugee camp in Jordan came from a Syrian refugee – but in most cases even this possibility is foreclosed. Just as the acousmatic voice is a voice physically separated from its source, so the unattributed texts are separated from their authors. Furthermore, since the whispering is purposively void of dramatic articulation (a manifestation of Pope.L's 'deadpan aesthetic'), the work smooths over whatever affective or personalizing inflections a textual fragment may possess. Lists of numbers and harrowing testimonies of war are recited with equal impassivity. In this respect, *Whispering Campaign* achieves multiple separations: of voice from source, of text from author, and of speech from affect.

Such affective discrepancy between words and their manifestations is a long-standing feature of Pope.L's art. An example is his *Skin Set Drawings*, the one other work of Pope.L's exhibited at documenta 14. Begun in 1997, the *Skin Set Drawings* now comprises several hundred works; at documenta eight were shown.[19] Each drawing is a stylized graphic rendering of a verbal statement that also serves as the work's title. The documenta selection included *Green People Are a Recent Invention* (2011), *Black People Are the Silence They Cannot Understand* (2001–2), *White People Are the Cliff and What Comes After* (2001–2), *Yellow People Are the Dog's Seed* (2010), and *Orange People Are the Way Things Used to Be When They Were in Power* (2012). The drawings vary in size and medium, but most were made with ordinary stationery items like marker pens and graph paper. The graphic techniques Pope.L uses in the *Drawings* vary drastically, from readily legible block capitals to mannered forms that render the letters virtually unreadable. In each case, Pope.L leaves the viewer to reconstruct the relation between graphic rendering and verbal statement (itself often confounding), and ultimately to derive significance from the reconstruction process itself. As Darby English writes of the *Skin Set Drawings*:

> To look at words written this way on a page, of course, is to find much to see in the body of a single letter – to register not only something of the word's weight but also the tension between its role in a verbal message and its visual condition of being-for-itself. Reading is not necessarily suspended, but by spacing out and interrupting it, these letters do disfigure it to an important degree.[20]

Why 'disfigure' reading in this way? According to English, the tension between reading and looking allegorizes the problem 'not of language as such, but of language as we use it to effect differentiation and negotiate differences'.[21] He goes on to distinguish between two ways in which language mediates difference. The first subjects human life to 'summary, description, measurement, assessment, an impulse that would sentence human singularity to a self-consistent repeatability that is, finally, unlivable in the extreme'.[22] Here, language effects difference by using terms pertaining to race, gender and other identifiers to shepherd people into established social categories. But English sees a different kind of difference at work in the *Skin Set Drawings*, one that 'abrades knowing, eats into settled conceptualizations, mauls them', rather than reconfirming established social categories.[23]

While English sees the latter kind of difference encoded within the *Skin Set Drawings*, the artist has recounted how his own intellectual preoccupation with ignorance (he identified the *Routledge Handbook of Ignorance Studies* as an important inspiration)[24] led him to develop *Whispering Campaign*:

> Sound works temporally. You only have the hearing-time someone is willing to give. And my awareness, as the instigator of the encounter, is always delimited by what I can know or offer at any one time. So – what is this about, this conjunction of time and knowledge? In what space do they meet? They meet in the space or architecture of ignorance. The construction of what we do not know. Or do not want to know. Or know but do not want to know. It's a constructed space made of mechanisms of obscuring, removal, disavowal, and disassociation. I realized that my pivot was ignorance and went on from there.[25]

While knowledge abrasion in the *Skin Set Drawings* takes place through limiting legibility, Pope.L locates ignorance at the limit of 'hearing-time'. It is an artifact of the transience of sound and of the speaking voice in particular. But despite the very different media in which they are realized, the *Skin Set Drawings* and *Whispering Campaign* share formal as well as conceptual characteristics. English describes the formal language of the *Skin Set Drawings* as 'spread' or 'sprawl'. He writes: 'A sprawl of letters, words, and possible significances retains a concept of semantic spread that proves fundamental for the project's management not of language as such, but of language as we use it to effect differentiation and negotiate differences.'[26] Put otherwise, each drawing comprises a scattering of signs that confounds the viewer's (or reader's) well-learned faculty of toggling between reading and looking to make sense of visual signs. *Whispering Campaign* uses a similar, though more literal, formal strategy of spread or sprawl. Whereas in the *Skin Set Drawings* semantic spread entailed confusing the viewer's faculties of reading and looking, the semantic content of *Whispering Campaign* was physically spread throughout Kassel and Athens by means of a network of speakers and other sound technologies. But as in the *Skin Set Drawings*, *Whispering Campaign*'s formal dispersion – in this case, in sound – can be connected with issues of political epistemology.

Like English, Dieter Roelstraete associates *Whispering Campaign*'s form with struggles over the meaning – social, cognitive, affective – of difference. Describing *Whispering Campaign* as a 'sonic scattering – an *infestation* of sound', Roelstraete connects his own vocabulary to the epidemiological terms within which Europe's migrant crisis has been couched:

> And I'm consciously marshaling the pathologizing language of certain
> political responses to the migrant crisis here – the old fear-mongering
> chestnut of xenophobic discourse: cast the alien, the foreigner, the other
> as an invasive species, as an infection or contagion. The work consisted
> of sound bites that seemed to spread like *viruses* . . . Diasporic or 'migrant
> sound', if you want.[27]

Furthermore, the whispered delivery emphasizes the work's proliferative form. As Roelstraete relates, 'the fact that [the texts] were whispered rather than shouted out made them feel all the more omnipresent and pervasive, *invasive* even'.[28] Shouting needn't have been louder than whispering in this case, since all the sounds in *Whispering Campaign* were amplified. Rather, I take Roelstraete to be implying that whereas people shout in order to maximize the immediate audience for their words, people whisper to minimize it, thereby allowing them to spread (mis)information without attracting public objection. Moreover, Roelstraete's comments suggest that the implications of semantic spread in *Whispering Campaign* are more ambivalent than in English's reading of the *Skin Set Drawings*. Spread or sprawl imitates not just the movement of migrants but also the language used by those who wish to pathologize the social difference that migrants signify. On an affective level, the uncanniness attributed to Pope.L's whispers shades into the fear and paranoia stoked by right-wing politicians at the time the work was conceived. Responses to the 'Ignorance is a virtue' slogan – such as those associating it with Donald Trump – further evidence the work's nexus of affective and political ambivalence.

In broaching the politics of knowing and unknowing, Pope.L's work also engaged with the theme of documenta 14's programming as a whole, which itself met with controversy. The exhibition's binding theme was pedagogy. By titling the exhibition 'Learning from Athens', the organizers intended to correct a bias in perspectives on the continent's ongoing financial and migration crises by substituting a northern European (i.e. German) standpoint for the view from 'the South' (i.e. Greece).[29] But this agenda did not insulate the exhibition from local resistance in Athens. Some Athenians saw the German art institution's presence as exacerbating, rather than repairing, the quasi-colonial dynamic between the two nations that began a decade earlier with Greece's highly politicized EU bailouts.[30] (Germany is the bloc's biggest economy and the bailouts were widely regarded as German handiwork.) Eleana Yalouri, a Greek anthropologist who, with Elpida Rikou, led the revanchist research project 'Learning from documenta', questioned what kind of 'learning' documenta's organizers might have in mind. 'We would like to understand better what documenta's arrival in Athens means, why in Athens and why now, and how are they going to go about it. . . . A lot has been written about the notion of learning and how knowledge is engaged with other forms of power.'[31] As if in response to Yalouri's concerns, head curator Adam Szymczyk declared at documenta's opening: 'We believe that unlearning everything we believe to know is the best beginning . . . the great lesson is that there are no lessons.'[32] But in making this statement, Szymczyk hardly divested himself or his exhibition of the role of pedagogue-at-large. One journalist sarcastically characterized him as a self-styled 'zen master'.[33]

While *Whispering Campaign* was realized under documenta's auspices, its presentation of pedagogical ethics both ventriloquizes and undermines what documenta officials thought 'Learning from Athens' could or should entail. Evidently, documenta's organizers thought *learning from* Athens necessitated *being in* Athens, and it was the very presence of one of the world's largest art exhibitions in their city that irked Greek critics like Yalouri as much as any aspect of its programming. But if *Whispering Campaign* was 'among documenta 14's most talismanic' works, it was so by virtue of warping documenta's agenda – however one interprets it – into an evanescent sprawl of anonymous, incorporeal and polyglot voices.[34] Pope.L's use of acousmatic techniques decoupled the whisperers from their whispers, which at once made them seem omnipresent and, as the whispers or their listeners travelled through the city, let them persistently fade away. By transposing the formal sprawl of his *Skin Set Drawings* from a graphic medium to a sonic one, Pope.L presents knowledge (or ignorance) as neither necessarily benevolent nor malevolent, but as a problem at the heart of encounters with difference. And through adapting his deadpan aesthetic to sound, Pope.L allows the political weight of *Whispering Campaign* to pivot on a conspicuous withdrawal of affect. Lauren Berlant identifies Pope.L's deadpan as a 'performance of withholding' that is at once amicable and aggressive, interpellating its audience into an 'absurd intimacy'.[35] But *Whispering Campaign* pushes this deadpan aesthetic to an extreme by using acousmatic whispering, removing even the 'pan' (slang for 'face') from the equation. *Whispering Campaign*'s use of acousmatic sound makes deadpan's 'performance of withholding' all the more literal; the disembodied nature of its solicitations makes its sense of intimacy all the more absurd. What the work performs is the withdrawal of voice from the voice.

Notes

1 Earlier in his career, Pope.L went by the name William Pope.L.
2 Darby English has written particularly compelling commentaries on race in Pope.L's work. See Darby English, *How to See a Work of Art in Total Darkness* (Cambridge, MA: The MIT Press, 2007) and Darby English, *To Describe a Life: Notes from the Intersection of Art and Race Terror* (New Haven and London: Yale University Press, 2019).
3 Pope.L and Dieter Roelstraete, *Pope.L: Campaign* (Milan: Mousse Publishing, 2019), 36.
4 Pope.L and Roelstraete, *Pope.L*, 97, 18.
5 Pope.L and Roelstraete, *Pope.L*, 83.
6 Pope.L and Roelstraete, *Pope.L*, 54–5.
7 Elana Schor, 'McCain Sends Out "Truth Squad" to Counter Whisper Campaign', *The Guardian*, 17 January 2008, https://www.theguardian.com/world/2008/jan/17/usa.uselections20081 (accessed 6 May 2020).
8 Pope.L and Roelstraete, *Pope.L*, 38.
9 A selection of these fragments was published on an A1 handout available to documenta visitors, where they are gathered under thematic headings including 'character portraits', 'calamities', 'spy texts' and 'geography collages'.
10 Lauren Berlant, 'Showing Up to Withhold: Pope.L's Deadpan Aesthetic', in *Showing Up to Withhold*, ed. William Pope.L and Karem Reimer (Chicago and London: University of Chicago Press, 2013), 110–11.
11 Each of the scholars discussed in this and the next paragraph offers an account of Pythagoras's acousmatic practice.
12 Schaeffer refers to the acousmatic sounds that are produced electronically as 'sound objects'. In his 1966 *Traité des objets musicaux*, Schaeffer aimed to give conceptual resolution to the novel mode of perception that sound objects inaugurated: 'Through listening to sound objects with their instrumental causes hidden, we come to forget about the latter and attend to the objects in their own right. The dissociation of sight and hearing encourages another way of listening: listening to sound forms, without any other aim than to hear them better, so that we can describe them through analyzing the content of our perceptions.' See Pierre Schaeffer, *Treatise on Music Objects: An Essay Across Disciplines*, trans. Christine North and John Dack (1966; Berkeley, CA: University of California Press, 2017), 66.
13 Mladen Dolar, *A Voice and Nothing More* (Cambridge, MA: The MIT Press, 2006), 61–2.
14 Rey Chow, 'Listening After "Acousmaticity": Notes on a Transdisciplinary Problematic', in *Sound Objects*, ed. James Steintrager and Rey Chow (Durham, NC: Duke University Press, 2019), 120.
15 Chow, 'Listening After "Acousmaticity"', 123.

16 'Radio, gramophone, tape-recorder, telephone: with the advent of the new media the acousmatic property of the voice became universal, and hence trivial. They all share their acousmatic nature, and in the early days of their introduction there was no shortage of stories about their uncanny effects, but these gradually waned as they became common, and hence banal'. Dolar, *A Voice*, 63.

17 Pope.L and Roelstraete, *Pope.L*, 84.

18 As Simon Frith has written: 'The microphone allow[s] us to hear people in ways that normally impl[y] intimacy – the whisper, the caress, the murmur.' See Simon Frith, *Performing Rites: On the Value of Popular Music* (Cambridge, MA: Harvard University Press, 1996), 187.

19 documenta 14, 'Pope.L'. https://www.documenta14.de/en/artists/13513/pope-l (accessed 4 May 2020).

20 English, *To Describe a Life*, 46.

21 English, *To Describe a Life*, 47.

22 English, *To Describe a Life*, 74–5.

23 English, *To Describe a Life*, 63–5.

24 See Megan O'Grady, 'Answering Society's Thorniest Questions, With Performance Art', *The New York Times Style Magazine*, 2 March 2018, https://www.nytimes.com/2018/03/02/t-magazine/pope-l-artist.html (accessed 21 May 2020).

25 Pope.L and Roelstraete, *Pope.L*, 13.

26 English, *To Describe a Life*, 47.

27 Pope.L and Roelstraete, *Pope.L*, 13.

28 Pope.L and Roelstraete, *Pope.L*, 13.

29 Cathryn Drake, 'Here's Why Greece Is Not Exactly Rolling Out the Red Carpet for documenta 14', *artnet news*, 29 March 2017, https://news.artnet.com/art-world/greece-not-rolling-out-red-carpet-documenta-14-905459 (accessed 21 May 2020).

30 Drake describes Greece as 'a de facto German colony'. See Drake, 'Here's Why Greece Is Not Exactly Rolling Out the Red Carpet for documenta 14'.

31 Drake, 'Here's Why Greece Is Not Exactly Rolling Out the Red Carpet for documenta 14'.

32 Hili Perlson, 'The Tao of Szymczyk: documenta 14 Curator Says to Understand His Show, Forget Everything You Know', *artnet news*, 6 April 2017, https://news.artnet.com/art-world/adam-szymczyk-press-conference-documenta-14-916991 (accessed 21 May 2020).

33 Perlson, 'The Tao of Szymczyk'.

34 Pope.L and Roelstraete, *Pope.L*, 12.

35 Berlant, 'Showing Up to Withhold', 110–11.

Sing

Katherine Nolan

Introduction: The sound affects of the power ballad

To sing is to 'make musical sounds with the voice, especially words with a set tune'.[1] This entry examines two video performances that conjure 'singing into a hairbrush' fantasies performed in public spaces. *Fight the Rising Odds* (2015) in which the singing is energetic, exercised and shouted against the wind as a form of 'letting it all out'; and *Breathless* (2017) in which the voice is choked, held back and almost silent as an act of 'holding it all in'. As a set they raise questions about singing and its relationship to emotion and affect, as well as challenging conceptualizations of the body through binaries such as interior/ exterior and public/private. Employing a practice-based methodology, I will analyse these artworks from the point of view of the body making sound/ affects and the listening/viewing body that witnesses this act (live or through the still-moving image). I aim to contest conceptualizations, such as Brian Massumi's, through which affects become emotions when their preconscious potential is sociolinguistically fixed.[2] I argue that this is problematic, from a poststructuralist point of view, in that it positions affect before language and posits a primal body before a cultural body. Through this practice-based methodology and poststructuralist approach, I aim to provoke and analyse a more complex interplay between language, the body and what I will now refer to as emotion/affect, at play in acts of singing.

There are many forms of singing across historical and cultural traditions, which may play different individual and collective functions, and may be performed formally or informally: from an Inuit mother's throat singing to calm the infant on her back to the Spirituals of African American enslaved communities or the mournful ballads of the Irish diaspora. This entry focuses on the act of singing as it relates to Western popular culture and specifically power ballads of the 1980s and early 1990s. I set out to examine the pleasures and pains offered by the genre's excessive emotionality and kitsch melodrama through performance art practice.

Routinely about pining for lost love, through characteristic affective 'power', the power ballad is characterized by the continual climb upwards in tempo, expressiveness, intensity and dynamics.[3] This entry particularly focuses on the power ballad as sung by female vocalists and the models of overemotional femininity such performances produce through voice, music and also dramatic bodily gestures. I argue that these popular culture songs are now inextricable from the imagery, visual and bodily language of their videos and associated performances, such that they become sound-images.[4] That is, to listen or to sing the popular culture song is an engagement not just with the auditory but also with the visual language of the body performed through the song.

The works I will discuss are part of the *Silent Video Series* in which I perform power ballads of the 1980s and early 1990s out of place in contemporary public contexts, a set of guerrilla performances. Through singing out of key with my earphones in against the background noise these artworks represent not only reperformances of these songs but parodic performances of myself as excessive, glamourized fantasies of femininity. I set out to deliberately invest in these melodramatic songs as a form of performance practice. That is, to knowingly attach my own emotional/affective states to these commercialized, cliché, predetermined accounts of women's emotionality as both an experiment and a deliberate provocation. Through this analysis I examine what these embodied experiences and readings of them reveal about singing, embodied sound and emotion/affect. I begin by discussing *Fight the Rising Odds* to draw out the cultural tropes of the power ballad and the mobilization of its sounds and affects in performance, and then discuss, in contrast, the anti-performance of singing in *Breathless* as a silent act.

'Letting it all out': *Fight the Rising Odds* (2015)

In 2015, I made a series of video performance artworks, which began as a response to feeling agitated by my personal situation at the time. I wanted to do and act and think with my body; to feel myself being propelled forward. On a grey day I rode my bike to the coast in Dublin. I positioned my camera, turned on my MP3 player and badly belted out Bonnie Tyler's 'Holding Out for a Hero' as passersby took their dogs for their daily walks.[5] What does this first gut reaction – which became a method of investigation – provoke about singing as a sound affect? With my ears plugged and the salt wind howling around me, I could less hear my voice than feel its effects on my body. Mentally returning to the scene now activates embodied memories: I feel the vibrations in my throat, the gulping inwards of breath to then push it back outwards with my diaphragm. As I write, I recite phrases of the song as a way to reposition myself back in that moment. I rehearse Tyler's forlorn pleas to be saved by clichéd, masculine heroes and gods.[6] As the notes move up and down the scale, I feel the sound vibrations change position in my throat. When the intensity of the song increases, I feel what I was seeking to experience on that day: an outward motion. Was I therefore 'letting it all out', that is, releasing emotions/affects from inside of me?

Eduardo Coutinho, Klaus R. Scherer and Nicola Dibben conceive of singing as externalizing emotions through the human voice, the encoding of which is determined through the 'emotional script' and the singer's affective state.[7] In this view, the emotional arc is predetermined and commingles with the singer's affects and is then externalized. Yet if one follows a predetermined emotional sequence, can it be said to be a release of one's emotions/affects? Both the act of singing and listening to songs are considered forms of catharsis. Catharsis can be defined as the 'purging' of excessive emotions.[8] For Aristotle, tragedy was cathartic as it allowed men [sic] to release 'undesirable emotions through vicarious experience, especially through seeing them represented on the stage'.[9] For Sigmund Freud, catharsis was a process by which the energy of repressed emotions/affects is discharged by being brought to consciousness.[10] Through such conceptualizations, in Western culture the energy of emotions/affects is dominantly considered to be 'releasable' from the body. In this entry, I will further interrogate this concept of the discharge of emotion/affects through song.

When I sing the power ballad's prescribed emotional script, the force of its steeply rising intensity begins to vibrate and affect my body. As I stand on the edge of the land, my earphones in, I am shouting outwards but also turning my attention inwards in order to perform the song. I recite and, at the same time, hear myself sing the words and melody. Sound is available at once as a lived embodied experience and perceptual encounter. Susanna Paasonen asserts a concept of affect as *resonance*, which occurs when subjects and vibrating bodies become attuned to and affect one another.[11] This term evokes affect simultaneously as both embodied and transmissible. The resonances of singing are not only an externalization to which others might become attuned; they are self-affective. Thus, in my act of singing, resonance might be the effect of me consciously pushing air outwards with my diaphragm which vibrates and warms my vocal chords or the ripples of sound as they pass through my throat and voice box; it might be the reverberations that amplify in the cavities of my chest or nose, the frequencies that pass through my body, or the associations or memories that are provoked by the sensations of sounds, tones, enunciations and lyrics;

and resonance may be the affects that these collectively provoke. As well as emotive sounds coming out of the body, emotive sounds resonate within the body as a form of self-affect.

Texts as affects and affects as texts

As I sing on that grey day, in the making of *Fight the Rising Odds*, I selectively read the lyrics through the lens of my own affective state: I wasn't, as Bonnie Tyler exclaimed, in need of a hero, but I did want to oppose the odds I faced. As I write, I reperform and listen to myself recite the song, the elongated 'herooooooo' becomes an exclamation, a yearning, a woman's lament. I ask myself, when I sing this song, what am I consciously engaging with and what am I subconsciously becoming through the gendered language of the song? I am performed through the song and its language, in the sense of both the lyrics and the embodied stylizations of gendered subjectivity[12] that the song (as sound-image) suggests. This reiteration of a stylized model of femininity becomes my lived, embodied affective experience. The relationship between language and affect is a contested one. For Massumi affects become emotions when their preconscious potential becomes sociolinguistically fixed.[13] In this view, language is therefore separated from affect. Eugenie Brinkema, however, asserts that '*Affect is not the place where something immediate and automatic and resistant takes place outside of language. The turning to affect in the humanities does not obliterate the problem of form and representation. Affect is not where reading is no longer needed.*'[14] Furthering Brinkema's critique, I seek to move away from theorizations of affect which polarize language and embodiment. I argue that the separation of affect and emotion via theorizations such as Massumi's produce problematic binaries of preconscious/conscious, primal/sociocultural, interior/exterior, felt/thought and posits a pre-discursive body that exists prior to language, culture and representation. Invoking Judith Butler's assertion that we can only know the body through language,[15] I seek to unravel a much more complex interplay between language, embodiment and emotion/affect. I consider in my analysis what happens when texts become affects and affects become texts, as well as asserting or rather performing textuality itself as affective.

Cycling once again through the musical and emotional script of the song, I recognize it as a rehearsal of embodied cultural significations that resonate in my body, and which I draw inwards as I swell my chest with my breath. The air warms and vibrates my throat as sounds are pushed outwards. As I enunciate and simultaneously listen to and feel my own voice and its vibrations, I also imagine myself imitating Bonnie Tyler's performance of the song in both voice and bodily gesture, as she invokes a desire to be heroically saved by a white knight on his trusty stallion or caught by superman as she helplessly swoons. Through the lyrics I am brought into a story world, in which the singer melodramatically laments a lost/yet-to-be-found lover. She is positioned as victim in this gendered narrative, establishing the male protagonist's power and sexual prowess against her passive longing. Claire Kahane cites the discursive production of gendered subjectivity through a dominant Western Oedipal narrative in which 'it is the masculine hero whose ambition, and desire drive the action, while the woman longs to be the object of his love'.[16] The words, pitch and tone, and the underlying pulse of the metre of the song I sing carry cultural significations and prescribed emotions that I imagine myself into: a dramatically passive feminized lament. What I enunciate are the sounds and embodiments of the tragic diva's cries. I experience a sense of declining pressure as I push air from my lungs with my diaphragm and then exhale a slow elongated sigh: 'hero ooohh'. Through embodied self-affect, I experience myself as a trope of feminized sorrow, crying out. The lyrics become my utterances, as I find myself stating that I wait to be saved by this ultra-masculine hero.[17] The stereotypical gendered narrative wants to speak me into a passive position waiting to be saved, which is clearly not the agency that I sought at the time. Rather than simply an externalization, the act of singing is also an internalization of the song and its sound affects, an ingestion and a pulling inward.

The recording of these performances as video works allows me to step into the role of reading these works as texts such that I become a body, reading my own body as sound-image, an audiovisual manifestation of the songs' performance. Outside of the act of singing as embodied experience, I listen to/view *Fight the Rising Odds*. I freeze frame the moments in which my body looks 'emotional' as a way to capture the intention of dramatizing what I was experiencing at the time. In these still images my head, face and shoulders arch backwards, my hair splays in a sweeping motion and my arm gestures dramatically. As my mouth gapes to let out the sound, it becomes a downturned cry, my brow furrowed. My movements stilled, I look pained, tortured, hysterically distressed. The act of singing the power ballad has produced me as an image of hysterical femininity.

Such dramatic contortions of the female body and face might be likened to the strange language of the hysteric's body.[18] Hysteria for Freud was repressed psychical trauma that returned as somatic symptom.[19] Thus, the hysterical body articulates what cannot be spoken such that the body itself can be read. Yet feminist analyses by Elaine Showalter, Georges Didi-Huberman and Linda Williams assert that the medical and psychoanalytic discourses that produced hysteria in the nineteenth century were driven by a desire to know and control women's bodies.[20] Through the emergent, yet dominant, positivist scientific paradigm, in which sight was privileged as purveyor of truth, that which was visible and readable on and through the body was deployed as proof of women's hysterical 'nature'. It was this context that produced what Showalter terms *The Female Malady*, an apparent epidemic of female hysteria.[21] Kahane argues that the threat that 'woman as speaking subject' posed to the social order of Victorian society, arising through the suffrage movement, led to a hystericization of women's voices.[22] The 'woman with a potent voice figured a kind of a vocal medusa both fascinating and fearsome'.[23] This trope of the wailing 'madwoman' still persists across cultural forms such as opera, literature, theatre, popular music and film through the enduring characterization of women as excessively emotional.[24] For instance, Susan McClary traces the trope of the madwoman as a frequent operatic narrative device (Lucia, Salome) and asserts how the 'repetitive, ornamental chromatic excess' of the singing is a marker of her 'dementedness'.[25] Such 'ecstatic excess' can be observed across cultural forms. In film, for example, Williams asserts how naturalized feminine hysteria is signified by 'cries of pleasure in porn, screams of fear in horror, and sobs of anguish in melodrama'.[26] The excessive sound affects and unstable tonality of the singing diva become somatic symptom and 'evidence' of her emotional volatility.

Feminist discourses have reclaimed historical accounts of hysteria, though not unproblematically: Showalter and Juliet Mitchell argue that this is exemplified in women's deployment of overemotionality as protest.[27] Katrin Horn asserts that the knowing performance of excessive femininity is 'an aesthetic strategy, which relies on parody – often achieved through stylistic exaggeration, excessive theatricality, or other forms of over-articulation – irony, and humour to create incongruities and discrepancies within (popular) texts'.[28] In the context of such strategies, consideration might be given to whether my performance was cathartic because I was 'letting it all out' or if it was instead a transgression of expected behaviours in this radical act of shrieking and shouting out of key on the side of a hill by the sea while people walked their dogs.

Singing without sound: *Breathless* (2017)

Fight the Rising Odds mobilizes 'letting it all out' as a form of transgression and parody. This chaotic excess is contrasted with the quiet 'holding it all in' of another work in the *Silent Video Series*. *Breathless* is also a performance of the melodrama of the power ballad, this time in the more bustling urban area of Dublin's inner-city Liberties. In the video, I walk into position in a black three-quarter-length dress, with sharp 1980s shoulder pads. Seagulls caw and vehicles pass as the screen fades up. As I begin to sing, no sound comes out. A faint tone can be heard from the headphones in my ears as I silently mouth the words. In this work, singing occurs with no sound. However, all the other aspects of the performance of the Western pop culture song are enacted: the gestures, movements and the emotive direct address to the viewer (see Plate 3). This attempt to hold in the 'letting out' (of sound, of emotion/ affect) creates a disjuncture which lends an absurdity and loneliness to the performance. The video dreamily moves between shots, as I look directly at the camera, gesture with my hands and gently sway to the rhythm of the unheard music. There is an 'out-of-placeness' that is accentuated by passersby going about their daily business. The sound of the singing, as the crucial form of resonance of the pop video form, is uncanny in its absence. Only the diegetic, atmospheric sounds of the streetscape – a passing car, a soaring gull and the tiny mummer of the song in the earphones – can be heard.

The song I 'sing' in this work, 'Show Me Heaven', was originally sung by Maria McKee. As I write, to recall my own performance, again I sing, rehearsing the lyrics and melody and noticing how they resonate in my body, and how they perform and affect me. I revoice the words sung by McKee: requests to be steadied amidst a state of quivering, fragile femininity. The register shifts upwards moving from breathy to more intense, shrill and soaring tones. As I sing, I find myself articulating a desire to be shown 'heaven', my vowel sounds – 'waaaay', 'oooooooohhhhh' – elongated, even orgasmic through the melody.[29] Sitting in my office, decontextualized, the singing becomes like a shrieking outburst, which breaks through the glamourized femininity of the original song: I shout an ecstatic, vulnerable femininity into the air, a hystericized femininity encoded as sexualized fragility.

Of course, in the video performance itself, no such sounds are actually emitted. The relative silence of the video artwork (all but) removes the emotional script of the song from audibility, yet the apparent effects and affects, as I listen to and perform the song, become legible in other ways. Though the singing is soundless, the emotional arc of the sound-image builds through the urgency of gestures: the hands reach out increasingly imploringly, the facial expressions become more strained and the pace of the video edit escalates. The singer's emotions/affects seem at once laid bare without the prescribed emotions of the song audibly dominating, and yet the prescribed gestural and audiovisual language of the video is further underscored. Through this conflict between the apparent emotions/affects of the singing and the underscored stylization of the performance, what is consciously 'put on' to perform this act and what is involuntarily felt through doing this act are muddied and called into question. Thus, through the relative silence, tensions between what might be acted and what might be felt arise. This points to the entanglement and inseparability of what is felt and thought, preconscious and conscious, autonomic and voluntary, and complicates such binaries which occur through the theoretical separation of emotion and affect.

There is a sense of 'holding in' such that both the sounds of singing and the emotions/affects that they signify appear to be choked back. Despite the silence, my throat is activated. Gasps and whispers escape, barely audible, such that emotions/affects seem to be partially revealed through what leaks through. As I silently mouth the words to the viewer, my eyes cloud up and become reddened, and a single tear rolls down my cheek. Tourist buses screech by and subtle gulping and guttural noises slip out, the sound affects of emoting while simultaneously holding back. The song increases in pace and climaxes. I collect the camera and walk away, as the video draws to a close. It appears that the tear 'escapes' as if revealing that which is being held back. Brinkema asserts that while the tear has been understood in Western culture as expressing a truth about interior states, since Aristotle, it has also 'been deployed to work through some of the most significant debates in philosophy about the relation between body and mind, the interior and exterior, the will and that which overrides will'.[30] Tracing how the tear has been variously conceived as the overspill of weakness or an act of purging (Homer), the instigator of sadness (William James), a muscular evolutionary response (Charles Darwin) and symptom in defence of unconscious fantasy (Freud), she asserts that the meaning of the tear is not self-evident but culturally inflected and demands to be read.[31]

As I revisit my performance as a video text, I pause it. As a still image of the face with the mouth gaping and brow furrowed, the eyes red and beginning to fill with tears, it reads as an image of sorrow. Laurel Nakadate's photographic series *365 Days: A Catalogue of Tears* (2011) interrogates how the photograph both mobilizes and problematizes the tear as if 'proof' of emotional pain. Still images of the artist both capture the fleeting moments of the water dropping and the face crumpling, and yet call the tear's truth into question through an implied conspiratorial relationship to the camera. While Nakadate questions the authenticity of her own acts of crying through their excess – crying every day for 365 days – the singular tear in *Breathless* can be read as an apparently authentic moment in the inauthenticity of the stylized, mimetic performance of the pop song. Yet there are other available interpretations that call this reading into question. Is the tear the force of affect silently escaping since the oesophageal route to the exterior is impassable? Or is it the act of clamping down and holding in itself that stimulates the single tear? Is the tear demanded or stimulated by the song? The tear comes, not in the traditional high climactic point of emotive intensity but in the lull of the second verse, an unremarkable point of non-tension or fervour. There is an apparent disjuncture between the arc of the song and the performer's signification of emotions/affects. This silent, falling droplet signifies an internal emotional trajectory that is discordant with the thrust of the songs implied narrative arc. Does the fact that this tear refuses to play by the rules of the power ballad's emotional script suggest that it is an unconnected internal stimulus, a memory or image unknown to the viewer? The tear, while punctuating the performance with observable emotions/affects visibly leaking from the body, is produced by, but also out of sync with, the song's rhythm, destabilizing the assumed connection between what we perceive as the singer's affects and the emotional script of the song. The act of 'holding in', and the 'radical quiet' or 'silence' that it produces, is an absence that amplifies this singular, silent, watery moment of apparent disclosure.[32]

Making the private public

The act of 'holding in' puts pressure on the interiority of the performer's body in its disjuncture with the readable emotiveness that is available to the viewer. The dominant Western understanding of emotions as being 'held in' or 'let out' is based on a conflation of the interior (especially of the body) with the private and the exterior with the public. Rebecca Schneider contends that a feminization of the private spheres of the home/family, made distinct from masculinized spheres of work/production, renders the bodies of women as emblematic of privacy itself: thus, the female body itself signifies the private.[33] Schneider's discussion denaturalizes the cultural concept of the private body that works to delineate the body's boundaries through sociocultural codes of propriety: prescribing what should be held back, hidden and kept private and what can be made public. Indeed, Brinkema challenges the very mapping of the body as interior through alternative spatial metaphors, such as Gershon's understanding of the bowel as part of the exterior world that runs through the body from mouth to anus and positing the oesophagus as an external rather than internal wall.[34] Thus, against those theories which position affect as secreted in the unknowable, unreachable, prelinguistic depths of a primal body, Brinkema refutes the interiority of the body itself. There is no interior body untouched by the public sphere: the interior/private body is itself a culturally produced concept.

Both *Breathless* and *Fight the Rising Odds* seek to complicate the dichotomization of the public/private, and the feminization of both interiority and emotionality, through strategies of disjuncture ('holding in') and excess ('letting out'). The positioning of this performance of 'holding it all in' in the act of 'letting it all out' in public is provocative. In the midst of the performance, I am both aware of and yet turned away from the public context. That is, I am aware of the act of making myself a spectacle, of passersby circulating around me and glancing as they go past. At the same time, I keep myself mentally apart, my focus inward. I connect with my internal landscape to provoke the performance but also maintain awareness of myself as image. While singing is often a public display attended as a social activity, the act of listening to songs on a personal device and 'letting it all out' is understood as a private act. The white earphones that stand out against the black dress signify the enclosed loop of solo listening not intended to leak into civic space. Within the video as text, the people who glance at me as they pass by emphasize the out-of-placeness of the performance and the making public of an act understood as private. These performances provoke an embarrassing inappropriateness; a twisting of the gut through the transgression of social codes. The work puts pressure on the concept of the internal body as untouched by the public sphere, by 'holding in' the 'letting out' in a public context in which it is illegitimated. The performance, as an act of transgression, both mobilizes and questions the splitting of the affective body into public and private bodies.

Conclusion

Analysing these artworks in order to interrogate the sound/affects of these reperformances of the power ballad, I read for the emotions/affects that the original songs signified as well as those provoked in me. I revisited the sounds I heard and felt in my own body, the ways that the act of singing effected and affected me, and the ways in which those acts signify and might be read. While seeking to interrogate the sound affects of the power ballad in this analysis, the currents have constantly moved me towards the visual: to external signs of emotions/affects that are readable through the body. The performance of these songs produced a visual embodied language. The act of singing impelled me to speak my body through the lyrical, melodic, rhythmic, gestural language of the power ballad and its gendered narratives. In these works, I reperformed 1980s and 1990s reincarnations of hysterical forms of femininity: Bonnie Tyler's smouldering and soaring belting tones, and Maria McKee's breathy, then glassy, almost shrieking vocals, were reperformed as transgressive and resistant. Sung out of place and out of key, the performances were both invested in, and in disjuncture with, the prescribed emotional and narrative arcs.

These artworks sought to examine and challenge the dominant Western conceptualization of singing as externalizing emotion/affects through alternate acts of 'letting it all out' and 'holding it all in'. The publicness of the outed private act of singing badly to the self is cringeworthy; the 'holding it all in' and miming of the song as a 'radical quiet' laid bare the performance as both stylized act and provocation of apparent emotions/affects. Understood through Butler's performativity of gender, there is no private, primal affective body before language or culture. These artworks complicate the binary underpinnings of theories such as Massumi's which separate emotion from affect. Furthering Brinkema's assertion of the textuality of affects, this analysis shows how affects emanating from the singer are not simply listened to, but also read through their visibility. This is amplified through the uncanny image of emotionality that does not let out any sound. Singing is a performance that traditionally follows a predetermined emotional cultural script as sounds designed to affect the audience and that are heard/read as the performer's own emotions/affects. The very act of making sound may affect the singer, its resonance reverberating within and beyond the body, yet the emotional script and the cultural significations of the song cause emotions/affects not to be 'released' but to be spoken through the body.

Notes

1 'Sing', *Lexico*, https://www.lexico.com/definition/sing (accessed 9 March 2022).
2 Brian Massumi, *Parables for the Virtual: Movement, Affect, Sensation* (Durham, NC and London: Duke University Press, 2002), 28.
3 David Metzer, 'The Power Ballad', *Popular Music* 31 (2012): 437–59.
4 After, but differentiated from, Deleuze's time-image, movement-image and affection-image. See Gilles Deleuze, *Cinema 2: The Time-Image*, trans. Hugh Tomlinson and Robert Galeta (1985; Minnesota: University of Minnesota Press, 1989).
5 Jim Steinman and Dean Pitchford, 'Holding Out for a Hero' (1984) [audio recording], UK: CBS Records.
6 Steinman and Pitchford, 'Holding Out for a Hero'.
7 Eduardo Coutinho, Klaus R. Scherer, and Nicola Dibben, 'Singing and Emotion', in *The Oxford Handbook of Singing*, ed. Graham F. Welch, David M. Howard and John Nix (Oxford: Oxford University Press, 2014), abstract text.
8 B. A. Farrell, 'Catharsis', in *The New Fontana Dictionary of Modern Thought*, ed. Alan Bullock and Stephen Trombley, 4th edn (1977; London: Harper Collins, 1999), 109.

9   Farrell, 'Catharsis', 109.
10 Farrell, 'Catharsis', 109.
11 Susanna Paasonen, *Carnal Resonance: Affect and On-line Pornography* (Cambridge, MA and London: The MIT Press, 2011), 16.
12 Judith Butler, *Gender Trouble* (1990; New York and London: Routledge, 1999), 11.
13 Massumi, *Parables for the Virtual*, 28.
14 Eugenie Brinkema, *The Forms of the Affects* (Durham, NC and London: Duke University Press, 2014), xiv.
15 Butler, *Gender Trouble*, 11.
16 Claire Kahane, *Passions of the Voice: Hysteria, Narrative and the Figure of the Speaking Woman 1850–1915* (Baltimore and London: The John Hopkins University Press, 1995), iv.
17 Steinman and Pitchford, 'Holding Out for a Hero'.
18 As seen in medical imagery of the nineteenth century such as Jean Martin Charcot's *Iconographie Photographique de la Salpêtrière* (1876–80).
19 Sigmund Freud, 'Frau Emmy von N, Case Histories from Studies on Hysteria', in Josef Breuer and Sigmund Freud's *The Standard Edition of the Complete Psychological Works of Sigmund Freud*, Volume II (1893-1895): *Studies on Hysteria*, 1893, trans. and ed. James Strachey (London: Hogarth Press, 1955), 86.
20 Elaine Showalter, *The Female Malady: Women, Madness, and English Culture, 1830–1980* (New York; London; Victoria; Ontario; Auckland: Penguin,1987; Pantheon Books 1985): Georges Didi-Huberman, *Invention of Hysteria: Charcot and the Photographic Iconography of the Salpêtrière*, trans. Alisa Hartz (1982; Cambridge, MA and London: The MIT Press, 2003); and Linda Williams, *Hardcore: Power, Pleasure and the Frenzy of the Visible* (Berkeley and Los Angeles: University of California Press, 1989).
21 Showalter, *The Female Malady*, 18.
22 Kahane, *Passions of the Voice*, ix.
23 Kahane, *Passions of the Voice*, x.
24 Susan McClary, *Feminine Endings: Music, Gender & Sexuality* (Minneapolis and London: University of Minnesota Press, 1991); Showalter, *The Female Malady*; Katrin Horn, *Women, Camp, and Popular Culture: Serious Excess* (Cham: Palgrave MacMillan, 2017); Christina Wald, *Hysteria Trauma and Melancholia: Performative Maladies in Anglophone Drama* (Basingstoke and New York: Palgrave MacMillan, 2007); and Mary Ann Doane, *The Desire to Desire: The Woman's Film of the 1940s* (Bloomington and Indianapolis: Indiana University Press, 1987).
25 McClary, *Feminine Endings*, 80.
26 Linda Williams, 'Film Bodies: Gender, Genre and Excess', *Film Quarterly* 44, no. 4 (Summer 1991): 4.
27 Showalter, *The Female Malady*, 147; Juliet Mitchell, *Women: The Longest Revolution: Essays in Feminism, Literature and Psychoanalysis* (London: Virago Press, 1984), 117.

28 Horn, *Women, Camp, and Popular Culture*, 21.
29 Maria McKee, Eric Rackin and Jay Rifkin, 'Show Me Heaven' (1990) [audio recording], UK: Epic.
30 Brinkema, *The Forms of the Affects*, 3.
31 Brinkema, *The Forms of the Affects*, 5–15.
32 Eugenie Brinkema, with composition by Evan Johnson, 'Critique of Silence', *Differences* 22, nos 2–3 (2011): 211–34.
33 Rebecca Schneider, *The Explicit Body in Performance* (London and New York: Routledge,1997), 72.
34 Brinkema, *The Forms of the Affects*, 150.

The disembodied voice

Julius Greve

In his short monograph on Kanye West's consensus masterpiece *My Beautiful Dark Twisted Fantasy*, Kirk Walker Graves includes a section that provides close readings of each individual track featured on the rapper-producer's 2010 album. In two of these readings, centred respectively on the songs 'Gorgeous' and 'Lost in the World' (and, in particular, the latter's second act, or coda, 'Who Will Survive in America?'), Graves utilizes a metaphor that signifies the use of voice in West's work and of the disembodied voice in particular. In the first instance, he states: 'The ambiguity of who is speaking in the hook – a type of figurative ventriloquism – is an important motif on *MBDTF* [short for *My Beautiful Dark Twisted Fantasy*]. Throughout the record Kanye indulges in certain impulses, moods, and reflections via the remove of a collaborator's voice.'[1] After several controversial appearances of the artist prior to the release of the album, above all at the 2009 MTV music video awards, the hook's lyrics of this second song on *MBDTF* 'serve as the complicated emotional confession of deepest turmoil,'[2] with the simultaneously self-aggrandizing and repentant speaker being rather humble on this particular occasion, conceding that he is on the verge of disaster, yet again: 'I can feel it slowly drifting away' and 'I will never ever let you live this down'.[3] However, it is his collaborator Kid Cudi who sings this part of the song, hence the claim that West's concession is realized by way of a 'figurative ventriloquism', a making-ambiguous of who is speaking or singing, who is the messenger or – to keep with Graves's metaphorics – the puppet and the puppeteer.

The second instance in which Graves utilizes his actually quite common ventriloquist metaphor is when he writes about how, on this characteristically sample-heavy album, the reappropriation of Gil Scott-Heron's politically charged 1970 poem 'Comment #1' – 'a scathing indictment of the student-led New Left's racial and historical naiveté' – leads to 'the album's final moments with the legitimacy of social prophecy'.[4] And, Graves adds, 'In a record devoted to the idea of redemption through excess, to the notion that an ambitious enough ego can translate the universe into a song cycle of the self, the choice of "Comment #1" succeeds. The poem as recast in "Who Will Survive in America?" is one such translation, Gil Scott-Heron's ghost ventriloquizing Kanye's angst and self-doubt.[5]

Cudi speaking for West, Scott-Heron's voice legitimizing West's 'redemption song' by means of 'ventriloquizing': the metaphor seems fitting to describe this particular Black artist and his musical, media-technological, cultural and, thus, racial positioning in the social ecology of celebrity in the United States. This is so, also because the metaphor has resurfaced more recently, after West had met the then president Donald J. Trump in the Oval Office on 11 October 2018. According to Michael Eric Dyson, what those who were watching were witnessing was 'white supremacy by ventriloquism. . . . A black mouth is moving, but white racist ideals are flowing from Kanye West's mouth. Kanye West is engaging in one of the most nefarious practices yet. A black body and brain are the warehouse for the articulation and expression of anti-black sentiment.'[6]

This entry, as replete with argumentative detours and digressions as it may seem, seeks to trace the maximally heterogeneous cultural and historical backdrop of the metaphorics of 'ventriloquizing' in order to show the relevance of the affectively charged sound of the disembodied voice. What is at stake in West's ventriloquism, I suggest, may be traced back to the Pythia of the Delphic oracle. The issue of ventriloquist speech also indexes debates within the humanities about what could arguably be called a vocal epistemology: an examination of the ways of knowing and expressing in and by one's own voice and those of others.[7]

Exchanging bodies

How to think disembodiment in relation to both sound studies and affect theory? What are the chances for thinking through the disembodied voice against the backdrop of both fields of research, given that these have, in the past few decades, homed in on what the naturalized phenomenology of Humberto Maturana and Francisco Varela had conceived of as embodied knowledge; as cognition that is actualized in and through the body? For Maturana and Varela, the system of cognition is 'a system whose organization defines a domain of interactions in which it can act with relevance to the maintenance of itself'; therefore, the embodied, lived experience of individuals is coterminous with processes of knowing, instead of the latter being defined by logical inference and representational mechanisms.[8] Put simply, '*Living systems are cognitive systems, and living as a process is a process of cognition.*'[9]

What is at issue here is not merely the correlation of the vital process
and that of knowing but the underlining of an ostensibly visceral level of
epistemology that has traction on how the question of sound and affect
may be posed within their respective fields of research. Indeed, the various
assumptions of major approaches to sound and affect within said fields have
equally been grounded in a downright somatic imperative, which itself had
emerged and established itself as a counterargument against the linguistic
turn within critical and cultural theory. Even though continental thought
of the past twenty years or so has not had a stake, for the most part, in what
Maturana had called the 'biology of cognition',[10] the epistemological upshot
of the concept of embodiment in the critique of representation points to
the same direction his and Varela's neurophenomenological critique of
mainstream analytical philosophy argued for: the conceptual sine qua non
of lived experience as a first principle. As Patricia Clough puts it: 'the turn
to affect did propose a substantive shift in that it returned critical theory
and cultural criticism to bodily matter which had been treated in terms of
... post-structuralism and deconstruction',[11] and the premise that materiality
is always already that of the signifier. For Clough, 'The turn to affect points
instead to a dynamism immanent to bodily matter and matter generally
– matter's capacity for self-organization.'[12] In this view, the linguistic turn
has, in terms of the rhetoric of intensity, affectivity and resonance, for some
time now given way to the championing of embodiment throughout major
strands of the humanities.

Certainly, this holds for sound studies and affect theory. In order to
think through the sounded and affective dimensions of lived experience, the
embodiment of cognition and expression seems an irremovable axiom vis-à-
vis the linkage of sound and affect. This is because similar to Maturana and
Varela's conceptual innovation of 'self-organization' in the reframing of the
process of cognition *as* the process of the living, in their repositioning of the
issue of representation within the production of knowledge, affect theory and
sound studies have each posited the corporeal nature of that-which-comes-
before or that-which-grounds or that-which-permeates either emotion or the
cultural formation of music.[13] In Steven Shaviro's helpful phrasing: 'Emotion
is affect captured by a subject, or tamed and reduced to the extent that it
becomes commensurate with that subject. Subjects are overwhelmed and
traversed by affect, but they *have* or *possess* their own emotions.'[14] What is more,
'The emphasis on embodiment, variation and relation' lends this approach
to affectivity 'an immediately political aspect', as Brian Massumi argues in
a 2008 interview[15] – an aspect that is underscored by Steve Goodman who,
in his book *Sonic Warfare: Sound, Affect, and the Ecology of Fear*, proposes
to examine 'the *politics of frequency*' displayed in 'the affective tendencies of

contemporary urban cultures' in terms of 'a nonrepresentational ontology of vibrational force'.[16] Such a politics supposedly precedes representational politics, while non-subjective immersion grounds subjectivities in cultural settings.

Christoph Cox, too, employs a somatic terminology in *Sonic Flux: Sound, Art, and Metaphysics*, when he discusses the vexed issue of 'phonography (literally, "voice-" or "sound-writing")'.[17] Preceded by a discussion of Jacques Derrida's account of writing as prosthetic and, thus, as resulting in a 'structural alienation from our bodily presence and animating intentions', Cox intimates:

> This is no less true of audio recording . . . As part of the archive of recorded sound, the recorded voice is submitted to the possibility of endless sampling, splicing, editing, and all manner of sonic modification. While it promises a return to the presence of the voice, audio recording does so at the price of an uncanny alienation of the voice from the body and mind that are said to have animated it. Recorded sound is often termed "disembodied," but that's not quite right. Rather, audio recording exchanges *one body for another*: a fleshy substrate for a mechanical one consisting of rotating plates, spinning discs, styluses, lasers, wheels, vibrating membranes and the likes.[18]

Why is the verdict of disembodiment 'not quite right'? Why is the process of recording sound and, in particular, that of recording voice, an instance of 'exchang[ing] *one body for another*' instead? Is it not precisely the case that the conception of encapsulating the voice mechanically or digitally, rather than in terms of 'flesh' and ostensible 'organicity', is premised upon the dissociation of one body's sound and/or voice from it, prior to another body's actualization of the latter? Granted, the linkage of sound and affect deals in corporeality. Subjects perceive or react to sound and affect, to frequency and pre-individual resonance on the level of the body prior to representational processes of making sense. Yet, such a corporeality may be assumed only on the basis of disembodiment – the instance of dissociation of sound and body, before sound may be perceived, felt or thought on an expressly 'visceral' level (the semantics of that predicate being helpfully recalled by Sianne Ngai as '"felt in or as if in the internal organs of the body"; "instinctive, unreasoning"; "dealing with crude or elemental emotions"', in short, as pointing to the domain of 'gut feeling'[19]).

The voice, too, is corporeally produced, each and every time: 'My voice comes from me first of all in a bodily sense. It is produced by means of my vocal apparatus – breath, larynx, teeth, tongue, palate, and lips',[20] Steven

Connor notes. And, yet, he adds, 'As a kind of projection, the voice allows me to withdraw or retract myself. This can make my voice a persona, a mask, or sounding screen.'[21] If the recording of the voice points to the alienation deconstruction ascribed to all kinds of writing, already the unrecorded voice uttered – the spoken word, the stammer, the stutter – is a constant reminder of the distance of the self from its other, of the self from itself, as psychoanalysis knows: for each embodiment a disembodiment, the latter, in turn, preceding the former as its condition of possibility. There is, however, another aspect to this alienating distance, effectuated as disembodiment in the spoken and recording voice: 'both hearing and emitting a voice present an excess, a surplus of authority on the one hand and a surplus of exposure on the other ... *One is too exposed to the voice and the voice exposes too much*, one incorporates and one expels too much.'[22] The excessive disposition of the voice is perhaps not grounded somatically. Rather, it is based on the fact that the voice leaves its body, its former source of origin, or site of generation, and that it might appear and reappear at a completely unexpected time and place: a vocal epistemology.

Pace Cox, and *pace* the understanding of what Massumi terms 'thinking affectively' as the event of 'body*ing*',[23] I want to underscore the notion of disembodiment (and in particular that of the voice) in the context of sound studies and affect theory. Building on Connor's seminal work *Dumbstruck: A Cultural History of Ventriloquism*, I will reconceptualize what it is to take seriously disembodiment as a media-specific mode of expression in literary and popular culture, from the entangled histories of ventriloquism, mysticism and blackface minstrelsy, all the way to *My Beautiful Dark Twisted Fantasy*. I will focus on the sound affect of the disembodied voice as it does not only haunt literary and media cultures, but the (post)phenomenological images of thought presupposed by major strands within both sound studies and affect theory.

The ventriloquial condition

The disembodied voice poses a problem for research on both affect and sound. If, as Cox argues, 'phonography ... exchanges one body for another',[24] what is the nature of the bodies aligned and compared, the one producing the sound recorded, the other recording and thus inscribing the former's sound on itself, so that it might be reproduced? Cox suggests a body made of flesh and one of mechanics. Yet, is the sound produced, or the voice recorded, itself corporeal? Connor's notion of the 'vocalic body' or 'voice-body' seems to offer a solution to the vexed problem of whether or not one should speak of the disembodiment of the voice in phonography or the production of further categories of corporeality: 'Voices are produced by bodies: but can also themselves produce bodies. The vocalic body is the idea – which can take the form of dream, fantasy, ideal, theological doctrine, or hallucination – of a surrogate or secondary body, a projection of a new way of having or being a body, formed and sustained out of the autonomous operations of the voice.'[25]

Connor's notion of the vocalic or sonorous body points to the way in which voices – recorded or not – have had a tendency of being partially independent of those bodies or objects containing them. The context in which this notion is proposed and developed, in order to address the dialectical relationship between processes of embodiment and disembodiment, is the history of ventriloquism, from the oracle at Delphi, via Charles Brockden Brown's early American republic novel *Wieland*, to Edgar Bergen's famous puppet Charlie McCarthy, all the way to contemporary recording technologies and their phonographic mechanisms. Put simply, what is at issue in the midst of the debates concerning sound, affect and (dis)embodiment is the agency, the autonomy of the voice, and the space and projected corporeality it commands and demands. For Connor, 'ventriloquism' is the description of the process or practice in which a voice (or sound) is dissociated from its putative source of origin or medium of generation and appears elsewhere. This understanding, as Connor mentions in his book, is strongly related to what Michel Chion has conceptualized as the 'acousmatic voice' that, in Mladen Dolar's words, 'is simply a voice whose source one cannot see, a voice whose origin cannot be identified, a voice one cannot place.'[26] Without going too much into detail concerning Chion's and, by extension, Dolar's theorization of the 'haunting effect' of 'the voice without a body',[27] in the present context we might think of the nature of the acousmatic voice, or of acousmatic sound more generally, as the counterpart of the more readily locatable and hence identifiable voice, in the same way in which Massumi differentiates between emotion and affect. A voice that cannot be placed is untamed and thus immersive; it is impersonal, pre-subjective and for that reason *haunting*. Moreover, as Connor notes, 'The experience of a voice without an obvious origin, whether in divine annunciation, oracular utterance, the voices of those seemingly possessed by spirits, or the many forms of auditory hallucination experienced by the psychotic and the ecstatic, is an experience of the overload of sound.'[28] Thinking through sound affects – the frequency of those emotive occurrences that cannot be possessed by subjects but that themselves possess subjects – ought to address precisely this cultural history of the overload of sound, a distortion of cultural semantics. In addition, what the acousmatic voice and the ventriloquist metaphor suggest is the always already excessive disposition of the voice, mentioned earlier, irrespective of whether it is spoken or recorded: '*One is too exposed to the voice and the voice exposes too much*.'[29]

To return to the concept of the 'voice-body', I suggest that what we are dealing with is only ostensibly a somatically grounded concept. Its naming is deceptive: in view of affect theory's axiom of embodiment and the cognitive processes it purports to explain, vocalic bodies are to be understood as the templates upon which, as Cox reasons, one body is exchanged for another in the process of audio recording. This exchange is premised upon the cultural history of auditory dissociation. Connor states: 'The history of ventriloquism is to be understood partly in terms of the repertoire of imaginings or incarnations it provides for these autonomous voice-bodies. It shows us clearly that human beings in many different cultural settings find the experience of a sourceless sound uncomfortable, and the experience of a sourceless voice intolerable.' Thus, the affective unpleasantness of these types of experience 'determines that a disembodied voice must be inhabited in a plausible body. It may then appear that the voice is subordinate to the body, when in fact the opposite is experientially the case; it is the voice which seems to colour and model its container.'[30] If the voice's autonomy is affectively intolerable, the dissociation from its source of origin marked as principally troublesome, does this problem also hold, on the level of discourse, for the apparent avoidance within affect theory and sound studies of taking seriously the disembodied grounding of what Massumi calls 'body*ing*'? Is the excess and disembodiment of the voice haunting not only as a sound affect but also as an issue to be tackled within the scholarly discourses of affect theory and sound studies? What may a cultural history of ventriloquism – of gut-talk and auditory dissociation – bring to light in terms of sound, affect and its connection against the backdrop of vocal embodiment and disembodiment?

Ventriloquism, Connor explains, has its practical and etymological roots in ancient Greece, with the *engastrimythos* being the one whose speech or story emerges in the stomach. The precursors of modern ventriloquists, the *engastrimuthoi* were known for being able to speak through the stomachs of other people.This practice was then later associated with the famed mystico-religious form of prophecy, for which the oracle at Delphi is known. As Connor notes, the rituals of Delphic prophecy, in which a divine voice would supposedly speak through the body of a priestess – the Pythia – was not linked merely to sounds coming from the stomach per se but generally to practices of divination, in which a voice was routinely dissociated from its origin. According to Connor, in this early form of ventriloquism, the auditory dissociation is that of an imagined deity, actualized by the body of the priestess.[31] Again, what is crucial is the presupposition of the vocalic body, the projected space of corporeality made possible by the voice in order for the voice itself to be actualized in the first place.

This divinatory practice is not only historically tied to a female actor, but it is also discursively marked as feminine, while later forms of ventriloquial speech would be framed along the lines of rationality, technicity and masculinity. With the eighteenth century figuring as a watershed of sorts, 'attention shifted away from the experience of the one possessed towards the powers of the ventriloquist who is believed to be capable of capturing others' voices though imitation and then "throwing" his or her (but almost always his) imitations away from himself and into others',[32] a shifting of focus from incarnation and figurative incubation to discarnation and vocal projection. 'In older conceptions of ventriloquism the body is the meeting point of the natural and the supernatural, the human and the inhuman; in modern ventriloquism, the body is acted on merely by the "natural magic" of the skilful ventriloquist.'[33] In both contexts, secular and non-secular, practices dominated by either male or female ventriloquists, the embodiment of one's voice is complemented and actually preceded by a disembodiment; a correlation that is essential for the traditions of both mysticism and, much later, transatlantic popular culture (and which is, arguably, replicated conceptually as the grounding of subjective emotion in the realm of presubjective affect, in the work of Massumi and others). What Eugenie Brinkema, in her admirable monograph *The Forms of the Affects*, has said of Gilles Deleuze's work (foundational as it is to affect theorists and sound scholars such as Massumi and Shaviro, Goodman and Cox) may equally be said of affect theory in general, since it, too, 'loses the subject only to hold tight to the body', catering, in the final analysis, to 'the potentiality of a visceral aesthetics'.[34] The ventriloquist problem – the issue of auditory dissociation in and for cultural history – arguably adds yet another layer to the succession from affectivity to subjectivity, from the realm of the body to that of particular bodies: a layer of conception which is itself not corporeally based but which enables the vocalic body to exist in the first place. The history of ventriloquism thereby merely underscores what is inherently at work in the phenomenon of the spoken or recorded voice, in its generative distancing from its adjacent corporealities as well as from itself, in each and every moment of its occurrence.

The co-extensive processes of embodiment and disembodiment also shed light on how religious discourse and popular culture become entwined especially in the American context. Leigh Eric Schmidt has critically examined the history of auditory culture and media apparatuses against the backdrop of the twin developments of the American Enlightenment and transatlantic mysticism, explaining how by the end of the nineteenth century already, the secularizing processes of technologies of hearing and speaking had significantly determined the 'shifting meanings of ventriloquism'.[35] Schmidt elaborates how 'The stage magician and his philosophical expositors made ventriloquism an easy and entertaining trick, a show of mastered simulation.' Thus, the newly transformed art of 'ventriloquism was indicative of the much larger absorption of religion into mediated, spectacular forms of modern consumption'.[36] The polyvocality indexed by this change from the sacred to the secular, while retaining portions of the former into the pop-cultural practices of the latter, also blurs the boundary between the gender ascriptions as per the incarnation or dissociation of voice. The correlation of the modern and ancient forms of ventriloquism, as evident in the inheritance of Delphic practices in what Schmidt calls 'oracles of reason' in the American Enlightenment and after, recapitulates the blurring of embodiment and disembodiment, as well as the culture of religious worship and that of popular culture.[37]

Along similar lines, Connor connects the mystical-religious (pre-) history of modern ventriloquism not only to the familiar practice in which the entertainer speaks to his audience through a puppet but also to the contemporaneous developments of the invention of new media and modern social relations and subjectivities. For here too, Connor argues, a given voice is separated from its site of generation, for example, by gramophone recording or by telephone and radio broadcasting, in the case of media apparatuses, and by the spatial architectonics and procedures of 'the talking cure':

> By the end of the nineteenth century . . . ventriloquism had come to know its place within popular culture. Ironically, actual ventriloquial performance becomes stabilized and standardized during the period in which ventriloquial relationships and scenarios become culturally generalized, in the figure of the mesmerizer and his subject, the psychoanalyst and his patient, the medium or the hysteric and her own body, the actor and the soundtrack, and in which a range of technologies for synthesizing speech make the ventriloquial condition a prevalent one.[38]

If modern ventriloquism of the eighteenth century had secularized the cultural connotations of being possessed or overwhelmed by a voice, the late nineteenth century and early twentieth century had shifted the singular aspects of that process of vocal dissociation into a heterogeneous array of cultural settings, from media apparatuses to psychoanalysis. Yet, while Connor argues that 'the meeting point of the natural and the supernatural, the human and the inhuman'[39] are the domain of older conceptions and practices of ventriloquism – Delphic, biblical and otherwise – the entwinement of these oppositional terms appears to be just as central when it comes to the modern and even contemporary actualizations of the vocalic body. Which is why the contemporary practice of speaking through the recorded voice of a collaborator (i.e. Cudi ventriloquizing West) may, or ought to, be linked to the very same history that made possible the vocal epistemology of Edgar Bergen's puppet and its theatrics of control and its loss: 'I can feel it slowly drifting away.'[40]

Of blackface and gut-talk

In recalling Kirk Graves's and Michael Dyson's descriptions of West's artistic work and political life in terms that are part of the cultural legacy I have just tried to contour, I want to suggest, finally, that the excess, the horror, the intolerable nature of the disembodied voice is culturally determined in a rather specific way, at least in the American context. This is because the notion of ventriloquism, routinely understood in the broad sense in which Connor denotes what I have also termed 'auditory dissociation', has appeared frequently in contemporary discourses concerning cultural and literary histories that are overtly racially marked (i.e. discourses that have significantly determined what the notion of 'race' signifies in the present).

Not merely the *engastrimuthoi* and the Pythia's glossolalia in ancient Greece,[41] nor the American Enlightenment and its debunking or 'unmasking [of] the divination techniques of antiquity',[42] nor Bergen's puppetry and phonographic media apparatuses, but the racial determination of voice in American history, arguably, lets the most impactful nineteenth-century form of popular culture in the United States – blackface minstrelsy – appear to be part and parcel of the larger, albeit discontinuous, history of ventriloquism. Indeed, Eric Lott claims: 'The blackface performer is in effect a perfect metaphor for one culture's ventriloquial self-expression through the art forms of someone else's' (somehow, anachronistically, a reversal of Dyson's logic according to which West merely figures as the puppet of white supremacy). And he adds that: 'What I mean to suggest about the character of popular culture in America is how unstable an entity it has been – a site of conflicting interests, appropriations, impersonations, indeed "nationalities," even in its allegedly national forms.'[43] What is at stake, then, besides religious and media-technological history is the appropriation of conflicting and commingling cultures and subcultures of slang and provocation via the disembodiment of voice.

The derogation of Black culture by the phenomenon of mostly non-Black actors and comics 'blacking up' as part of short theatrical acts and skits points to the uneasy relationship of American culture to its racially marked vocalic bodies. The disembodied voice, and the history of ventriloquism, is thus characterized not only along the heterogeneous categories of religiosity and secularity, femininity and masculinity, and corporeality and technicity but also and in terms of ethnicity and race. In all of these contexts, the vocalic body – 'a body-in-invention, an impossible, imaginary body'[44] – is of central importance, yet it includes the processes of both embodiment and disembodiment. Cognition is embodied, as is the affective set-up of contemporary social, ethical and aesthetic interaction, according to major strands within affect theory. And yet, the sound affect of the disembodied voice and its both spiritual and profane, literary and artistic, gendered and racially marked, as well as media-technologically determined history of ventriloquism suggest that embodiment presupposes dissociation, bodies require voices, and not vice versa.

The racially determined aspect of ventriloquism's cultural dynamics, therefore, provides an ethico-political answer to the question of why the voice without a body has consistently been described as haunting: this incorporeal voice, giving rise, nonetheless, to vocalic bodies, is the corollary of hundreds and thousands of bodies without voices, many of them having been rendered voiceless in the course of the Middle Passage and, subsequently, by means of blackfacing practices.[45] Such an answer to the question concerning acousmatic haunting indexes the political effectivity not only of processes of bodying and affective immersion but of auditory dissociation and the alienating potentials of the voice: a deafening silence. The excess of the disembodied voice mirrors the remainder of countless, voiceless bodies in the course of modernity, an equation that the whole of West's controversial work and his personae attest to. In effect, this is the cultural backdrop against which Graves is able to read crucial moments of *My Beautiful Dark Twisted Fantasy* as acts of ventriloquism.

Notes

1 Kirk Walker Graves, *Kanye West's My Beautiful Dark Twisted Fantasy* (London: Bloomsbury, 2014), 61.
2 Graves, *Kanye West's My Beautiful Dark Twisted Fantasy*, 62.
3 West cited in Graves, *Kanye West's My Beautiful Dark Twisted Fantasy*, 61.
4 Graves, *Kanye West's My Beautiful Dark Twisted Fantasy*, 121.
5 Graves, *Kanye West's My Beautiful Dark Twisted Fantasy*, 121.

6   Douglas Ernst, 'Michael Eric Dyson: Kanye West's Support of Trump Is "White Supremacy by Ventriloquism"', *The Washington Times*, 12 October 2018, https://www.washingtontimes.com/news/2018/oct/12/michael-eric -dyson-kanye-wests- support-of-trump-is/ (accessed 9 December 2021).

7   This notion is comparable to, yet not identical, with the more general conception of that term in Nina S. Eidsheim and Katherine Meizel's 'Introduction' to their *Oxford Handbook of Voice Studies* (Oxford: Oxford University Press, 2019), xx. See also the recent volume *Media Ventriloquism: How Audiovisual Technologies Transform the Voice-Body Relationship*, ed. Jaimie Baron, Jennifer Fleeger, and Shannon Wong Lerner (New York: Oxford University Press, 2021) and the way the editors theorize the titular notion: 'By acknowledging *media* ventriloquism as part of our everyday relationship with technology, we also recognize its social impact and the need to further examine the politics and ethics of separating bodies from voices' (1).

8   Humberto R. Maturana, 'Biology of Cognition', in *Autopoiesis and Cognition: The Realization of the Living*, ed. Humberto R. Maturana and Francisco J. Varela (Dordrecht: D. Reidel Publishing Company, 1980), 13.

9   Maturana, 'Biology of Cognition', 13. Varela writes elsewhere that in his approach, which has also variably been called enactivism and neurophenomenology, 'the view of cognition is not that of solving problems through representations, but as a creative bringing forth of a world.' See Francisco J. Varela, 'Whither Perceptual Meaning?', in *Understanding Origins: Contemporary Views on the Origin of Life, Mind, and Society*, ed. Francisco J. Varela and Jean-Pierre Dupuy (Dordrecht: Kluwer Academic Publishers, 1992), 255.

10 Maturana, 'Biology of Cognition', 2.

11 Patricia T. Clough, 'The Affective Turn: Political Economy, Biomedia and Bodies', *Theory, Culture & Society* 25, no. 1 (2008): 1.

12 Clough, 'The Affective Turn', 1.

13 Consider Brian Massumi's seminal argument, in *Parables for the Virtual*, that while '[a]n emotion is a subjective content, the sociolinguistic fixing of the quality of an experience', the order of 'the irreducibly bodily and autonomic nature of affect' somehow *precedes* that 'personal' content of lived experience. See Brian Massumi, *Parables for the Virtual: Movement, Affect, Sensation* (Durham, NC and London: Duke University Press, 2002), 27.

14 Steven Shaviro, *Post-Cinematic Affect* (Winchester: Zero Books, 2010), 3.

15 Brian Massumi, *Politics of Affect* (Cambridge: Polity Press, 2015), 51.

16 Steve Goodman, *Sonic Warfare: Sound, Affect, and the Ecology of Fear* (Cambridge, MA and London: The MIT Press, 2012), xv.

17 Christoph Cox, *Sonic Flux: Sound, Art, and Metaphysics* (Chicago: The University of Chicago Press, 2018), 77.

18 Cox, *Sonic Flux*, 77; emphasis added.

19 Sianne Ngai, 'Visceral Abstractions', *GLQ: A Journal of Lesbian and Gay Studies* 21, no. 1 (2015): 33.

20 Steven Connor, *Dumbstruck: A Cultural History of Ventriloquism* (Oxford: Oxford University Press, 2000), 3.

21 Connor, *Dumbstruck*, 5.
22 Mladen Dolar, *A Voice and Nothing More* (Cambridge, MA: The MIT Press, 2006), 81.
23 Massumi, *Politics of Affect*, 202–3.
24 Cox, *Sonic Flux*, 77.
25 Connor, *Dumbstruck*, 35. Throughout the present text, I will use the words 'vocalic' and 'vocal' nearly synonymously, though tending towards the latter – for instance in my outlining of a vocal epistemology. Whenever I use Connor's concept of the 'vocalic body' I will stick to his phrasing, as it differs from Adriana Cavarero's notion of the 'vocal body', in Cavarero, 'The Vocal Body', *Qui Parle: Critical Humanities and Social Sciences* 21, no. 1 (2012): 71–83. In this context, see also Baron et al.'s more recent conception of 'technovocalic bodies', that is, 'deliberate voice/image combinations that depend on technology for their existence' ('Introduction: Theorizing Media Ventriloquism', 2).
26 Dolar, *A Voice and Nothing More*, 60.
27 Dolar, *A Voice and Nothing More*, 61.
28 Connor, *Dumbstruck*, 23–4.
29 Dolar, *A Voice and Nothing More*, 81.
30 Connor, *Dumbstruck*, 35.
31 Connor, *Dumbstruck*, 49–52.
32 Connor, *Dumbstruck*, 197.
33 Connor, *Dumbstruck*, 197.
34 Eugenie Brinkema, *The Forms of the Affects* (Durham, NC and London: Duke University Press, 2014), 24–5.
35 Leigh Eric Schmidt, *Hearing Things: Religion, Illusion, and the American Enlightenment* (Cambridge: Harvard University Press, 2000), 9.
36 Schmidt, *Hearing Things*, 163.
37 Schmidt, *Hearing Things*, 78–134.
38 Connor, *Dumbstruck*, 398.
39 Connor, *Dumbstruck*, 197.
40 West cited in Graves, *Kanye West's My Beautiful Dark Twisted Fantasy*, 61.
41 Agnès Gayraud, 'Glossolalia/Xenoglossia', in *Unsound: Undead*, ed. Steve Goodman, Toby Heys and Eleni Ikoniadou (Falmouth: Urbanomic, 2019), 61.
42 Schmidt, *Hearing Things*, 81.
43 Eric Lott, *Love & Theft: Blackface Minstrelsy and the American Working Class* (Oxford: Oxford University Press, 2013), 95.
44 Connor, *Dumbstruck*, 35.
45 See Michael North, *The Dialect of Modernism: Race, Language, and Twentieth-Century Literature* (Oxford: Oxford University Press, 1998), 9, for an analysis of 'racial ventriloquism' in the self-dramatizations of American modernist authors. See also Mita Banerjee, *Ethnic Ventriloquism: Literary Minstrelsy in Nineteenth-Century American Literature* (Heidelberg: Winter, 2008), for a reading of 'ethnic ventriloquism' to be witnessed in the racial politics of major nineteenth-century American fiction.

Part III

Threshold sounds

Entry 9

Tin/ny

Rob Garbutt

I

From a survey of definitions in the *Oxford English Dictionary*, it is apparent that in received English – that is, the language received from the colonial metropole in contrast to the Australian vernacular – tin and tinny, the metal and its affective sonification, have accumulated negative connotations.[1] This derives from late nineteenth-century comparisons between common, inexpensive base metals such as tin and the precious metal silver.[2] Such comparisons were made, for example, between kitchenware made from tin-plate and silver or silver-plated steel, yielding the sense of tin as cheap or counterfeit. With increasing availability of recorded sound from the late 1800s, this deficit transferred smoothly to tinny as sound that lacks bass.[3] Later, in the mid-twentieth century, is yet another move from sound production to reception in the guise of the person with the 'tin ear' who is somewhat tone deaf and, in a more figurative sense, cannot grasp the core message in what is being said to them.[4] Tinniness, then, is an effect that is not generally pursued in sound production, except for those situations in which a thin, metallic sound might 'add value' to visuals already redolent with everyday commonality: the auditory cliché of muzak over the tannoy at a suburban mall comes to mind.[5]

In the Australian cultural landscape tin and tinny have multiple resonances and, unlike received connotations, it is not unusual for these to be positively alloyed with comfort, nostalgia or recreation. For example, a tinny is a small, sparsely outfitted, lightweight boat with an aluminium hull. With the addition of a low-horsepower outboard motor, the tinny is a practical and affordable craft for boating enthusiasts from many walks of life. In Australian culture, which espouses an egalitarian, even classless, ethos and where practicality trumps aesthetics, practical and affordable is good. On the weekend, waterways are abuzz with tinnies. Those aboard, though not the skipper of course, might add to their enjoyment by 'sinking a tinny or two', that is drink a few cans of beer. This meaning develops from tin referring to a metal-drinking vessel.[6] Thus, in the late 1970s, when the innovative Darwin-based architectural firm Troppo began researching early twentieth-century vernacular architecture in order to inform their own low cost, environmentally appropriate house designs for the Australian tropics, it was an obvious shorthand to group structures that used corrugated metal roofs under the category of 'tinnies'.[7] In each of these Australian instances, the association of tinny with notions such as affordability, practicality and 'common' taste does not immediately imply associations with negative affect.

While each of these Australian 'tinnies' has a sound that activates a range of affects – the lolling, aqueous slurps of estuarine waters on the hull of the tinny or the crack and phsh of opening a beer – when it comes to tin in the Australian *landscape*, it is the resonances of tin in the form of corrugated metal sheeting that are culturally dominant. While corrugated iron features strongly in the vernacular architecture of many nations, 'the factor that sets Australia apart is the predominantly positive light in which it is perceived'.[8] Architectural historian, Philip Drew argues that to the Australian anti-authoritarian and egalitarian sensibility, this positivity emerges from tin's 'oppos[ition] to conventional standards of civilisation typified by masonry architecture'.[9] There is, however, more to the warmth with which tin is held, and the tin roof in particular, than architectural anti-authoritarianism.

Corrugated iron was invented in 1829 by Henry Palmer 'in response to a desperate need on the British and London Docks for storage space'.[10] By corrugating iron sheets, Palmer added stabilizing arches into tough but bendable sheet metal. With this new material he could rapidly erect lightweight structures to accommodate the increasingly abundant goods flowing into London from the expanding British Empire. Due to the ready corrosion of iron, its durability came into question; however, this was answered with the discovery of the galvanizing process in 1837 and from then on corrugated galvanized iron quickly became a standard construction material, especially for industrial buildings.

In Australia, the usefulness of the material was readily appreciated. It was low cost, lightweight, easily transported, durable and, unlike tiles and shingles, did not require specialist construction skills. Moreover, in a land in which local building materials were often hard to find in quantity, particularly in the arid interior of the continent, tin became the first choice for durable structures. As Peter Myers writes, 'corrugated galvanised iron . . . formed the crucial link in the economic utilisation of the otherwise inaccessible physical environments of Australia'.[11] He continues, 'the wide availability of corrugated galvanised iron water tanks would alone explain the *ad hoc* dispersal of small social units across the Australian terrain'.[12] In this sense, colonial expansion and dispossession of Indigenous Australians were enabled by corrugated iron, and its use in vernacular architecture in rural areas became bound up with developing ideas of 'the bush' as core to Australian identity. In the 1950s, when the question 'What does it mean to be Australian?' became more urgent, Adam Mornement and Simon Holloway conclude that '[w]ith the benefit of hindsight it seems that corrugated iron had a role to play in addressing this question.'[13]

This entanglement of tin with Australian history and identity is accompanied by an affective sonic landscape. The remainder of this entry follows this affective and material entanglement by tracing two threads: first, the sonic activation of tin by elements of the Australian environment and culture; and second, by way of Jacques Lacan's concept of extimacy,[14] a neologism which in this instance expresses how these tinny sounds create affects that are simultaneously *external* (or public) and in*timate*. Two contrasting case studies will focus this examination; however, at this point, it is worth noting that neither case references the most common association between the environment, tin and sound: rain on a tin roof. D. H. Lawrence observed this cultural connection during his three-month visit to Australia in 1922 and referenced it in his novel *Kangaroo* where one of his characters, a First World War officer, reminisces, 'Oh, how I liked the rain on the tin roofs of the huts at the war. It reminded me of Australia.'[15] Australian country musician James Blundell explains this connection and affect in comments on his 1992 song 'Rain on a Tin Roof': 'The sound of rain on a tin roof and the smell of dry earth when rain starts, are two of the most stimulating sensory experiences available to the human being . . . [It] is the literal translation of comfort and security.'[16]

If rain on a tin roof is a source of positive affect, more ambivalent affects are associated with the way tin roofs respond to wind and sun. Glen Phillips situates his poem about summer, 'February', in an inland mining town where 'By November it was already hot' and the 'heat shimmered off corrugated iron roofs./The fences were mostly of the same tin/and threw back that searing sun straight/ into the windows of the crouching homes.'[17] The muffled inaudibility of a day baked into lethargy, before which homes crouch, is a common Australian summer experience. A more unsettling staccato ambience is deployed by Indigenous author Kim Scott in his novel *Taboo*: 'a sheet of corrugated iron clanged in the wind. A door knocked in its frame. A window rattled.'[18] These few examples provide a sense of how the activation of tin by rain, sun and wind can produce effects and affects that resonate with audiences because they form part of an Australian language of ambient sound. In each, sound co-creates personal experience and shared affect.

Two recent cultural productions enable a closer examination of tin's sound affects and position these affects in an Australian historical and cultural context. By sound affect, I am referring to how the sound of a sounding object circulates feeling and thereby galvanizes people into a public, even if this public is infused with an atmosphere of separation rather than connection.[19] As already noted, galvanizing has close connections with tin, and following Jane Bennett's lead, it also signals that sounding objects are members and co-producers of this excitable, assembled public. Chemistry, in its material and affective senses, is in the field of play. To consider the relations between the intimate inner world of affect and the outer public world, Lacan's concept of extimacy is useful. Extimacy has the topology of a Möbius strip or torus, in which a single surface disrupts the inside/outside binary. As Jacques-Alain Miller puts it, 'this expression "extimacy" is necessary in order to escape the common ravings about a psychism supposedly located in a bipartition between interior and exterior.'[20] Tony Hughes d'Aeth further explains how this neologism 'describe[s] the intimate exterior of psychic reality, a reality that transects the formal distinctions between public and private that organise social life'.[21] In this entry, this extimate affect of sound is situated in a specific aspect of the contemporary Australian political milieu, in which white settler Australians struggle with the facts and emotions of the nation being founded on a colonial land grab, as well as the consequent effects of this dispossession on Aboriginal Australians and Torres Strait Islanders. The struggle is one that often ignores and otherwise silences what will not go away. The presence in the polity of other non-Indigenous Australians is not denied here; however, it is a particular field marked by whiteness and settler colonialism that this entry opens for examination. In the two case studies that follow, tracing affects through the sonic activation of tin and Lacan's conception of extimacy enables the construction of assemblages in which the subject of Australian history, loosely described as 'Australian identity', becomes articulated with the materiality of tin via tin's 'sounding'.

The first case study examines the initial sequence of the twelve-minute 'Tin Symphony' section from the Opening Ceremony of the Sydney 2000 Olympic Games, a sequence that demonstrates 'corrugated iron's significance to the [Australian] national sense of self'.[22] Upbeat and optimistic, 'Tin Symphony' forms the developmental bridge between Opening Ceremony segments depicting uncolonized First Nations' Countries and the 'Arrivals' segment that depicts mid to late twentieth-century multicultural Australia.[23] This production, designed to tell a mainstream history of Australia, is contrasted with the second case study which focuses on a court scene from *Sweet Country* (2018) by Indigenous film-maker Warwick Thornton. This film, set in post–First World War central Australia, faces squarely the brutality inherent in the process of colonization, and it achieves this with a soundtrack that includes no film music, only dialogue and environmental sound, and in which tinniness plays a fundamental role.

II

'Tin Symphony' is a discrete section in the 'cultural segment' of the Opening Ceremony of the Sydney 2000 Olympic Games. Cultural segments typically interpret some aspect of the host nation's history, culture and achievements for a national and international audience.[24] The Sydney Olympic Games cultural segment had a planned duration of sixty minutes and had seven sections in all. At twelve minutes, 'Tin Symphony' (the huge stadium screen also shows the French translation 'Symphonie en tôle ondulée' or 'Corrugated iron symphony') was the longest section, occupying a pivotal position in what can be viewed as a national creation narrative. This narrative began with a primordial ocean (the 'Ocean Dreaming' section) out of which Indigenous Australian participants sang and danced Country into being ('Awakening') with its fire-adapted fauna and flora ('Fire' and 'Nature'). 'Tin Symphony' then surveyed Australian history from the first British sighting of the east coast of Australia in 1770 until the post–Second World War period. The 'Arrivals' section followed, depicting the waves of non-British migration that created contemporary 'multicultural' Australia. The cultural segment concluded with 'Eternity', a view to the future of Australia as a 'family of children . . . from many lands . . . whose time is just beginning'.[25]

And so, to the 'Tin Symphony' itself.

Into an orchestral accompaniment to Australian 'Nature' enters a British captain surveying the land and its people atop his tandem tricycle-cum-sailing-ship that is powered by two naval officers and one seaman below deck. The vessel halts, and with the sounds of the Sydney Symphony Orchestra ebbing, the captain raises a metaphorical flag by lighting a Roman candle. Its loud report echoes around the stadium. The wail of an electric guitar takes up this intrusion into peaceful 'Nature' and, moving into a fast-fingered solo with a martial beat of percussion, drives a horde of Sidney-Nolan-styled Ned Kelly figures into the arena, each holding a gun spewing a fountain of golden sparks. Into their midst rolls a monstrous fire-breathing mechanical horse on wheels which appears to be powered by a corrugated iron water tank with a huge exhaust pipe. Its body and neck form a derrick, while its head is shrouded in sheet metal. As it rolls forward, a brief eight-second reprise of Aboriginal music from the 'Awaken' section is broken by two cracks of a stock whip and the neigh of a horse in alarmed response. The soundtrack then straightaway breaks into Ian Cooper's 'Tin Symphony' theme, a folk-rock Irish jig, with a spotlight picking out the fiddler who is perched on the iron horse's neck.[26] With those two cracks of the whip, the violence of the frontier is over. The 'Tin Symphony' section can now be propelled onwards in dance. True to its title, tin plays a significant role in the segment. It features in tin water tanks rolling across the arena floor, while sheets of corrugated iron are used first as a dancing prop, then as a base for a tap dance sequence (see Plate 4). After a musical change in mood – a symphonic interlude – some Bluegrass finally arranges the corrugated iron to become houses, sheds and outdoor dunnies (an informal Australian term for a toilet). From one central shed, a shearing shed, animated bales of hay emerge – now to a carnival-like fairground tune – and are 'shorn', before splitting open to release a band of men in Mambo-style shirts each pushing a Victa lawn mower. In choreographed formation they create Olympic Rings, perhaps a reference to Australia's previous hosting of the Olympic Games in 1956 in Melbourne. The segment ends having spanned Australian history from British 'discovery' in 1770, invasion and colonization from 1788 to 1900, Federation of the colonies in 1901, to the apex of the white Australia Policy in the mid-1950s.

It is somewhat astonishing, though from the previous discussion it is also apposite, that tin should be the basis for telling the story of this complex period of Australian history from 1770 until around 1956. The opening portion of the section features Ned Kelly, a bush ranger of Irish cultural heritage who was hung for his crimes in 1880, aged twenty-five. Ned Kelly became famous for his resistance to the English-dominated imperial establishment and his use of iron armour. In the words of the official media guide to the Games, Kelly was 'an Australian "Robin Hood"' who robbed the rich to give to the poor.[27] In 'Tin Symphony', he is symbolically realized as a mob swarming across the landscape with fire-spouting rifle in hand and in the visually simplified iconic form that Sidney Nolan created for a series of paintings produced between 1946 and 1947. The painting shown on the stadium screen features Ned on horseback with rifle in hand riding away from the viewer into mallee country.[28] We see through the eye-slit in Ned's 'great rectangular metal helmet' blue sky. It sits above an expansive landscape, and both, according to the segment's director Nigel Jamieson, create 'a metaphor for the shuttered windows of the suburban houses and gardens from which the [non-Indigenous] Australians peeped out on the vastness of the continent'.[29] Iron, corrugated and otherwise, provides the material for the scene as Ned clears the way to install an egalitarian tin culture.

In this opening movement of 'Tin Symphony', the Aboriginal song from the 'Awakening' segment is abruptly overwhelmed with the amplified sound and urgent, lively beat of an eponymously named Irish jig. Appropriately enough, it is not elements of the Australian landscape that activates the corrugated iron. Instead, the jig is both the activator of tin and its resonance, that is an activation that arises from another land, Ireland, with its green, watered landscape: it is a nostalgic landscape for colonizers compared with the dry mallee that Ned faces. The energetic, joyous intensity of the jig that could be rain on a tin roof has the Opening Ceremony crowd clapping in unison in 'a celebration of Australian "larrikinism", and the energy, humour and ingenuity, that powered the settlement of the bush'.[30] In the later movements of 'Tin Symphony', the soundtrack transforms through Bluegrass to a final fairground carnival feel. Ned dancing across the mallee with a gun in his hand is just the beginning of the carnivalesque atmosphere, carnivalesque in the sense that Bakhtin characterizes as a temporary overturning of the established order. However, this carnival that overturns Aboriginal lore and institutes a dominant British order has not been temporary.[31] And this carnival of 'Tin Symphony' is doubly overturned because it is not the British overseers depicted as the colonizing force. Instead, colonization is displaced onto a band of Ned Kellys animated by an Irish jig: an Irish Catholic underclass whose labour fashions the larrikin, corrugated iron colony. While factually resonant, this is also a cover-up. The affect of the scene would be substantially changed if performed by mounted police moving to an English quadrille.

In the foot-tapping frolic of 'Tin Symphony', then, Indigenous Australians and the Australian landscape are rendered inaudible by a soundtrack from the Northern Hemisphere that activates tin and makes it resonant. This resonance, however, revels in the other of colonial settlement, the Irish tin to the English silverware. This national group that, in comparison to the English, is not the ruling elite in Australian history from 1770 to the 1950s, becomes extimate. That is, in carnivalesque fashion, in the global spectacle of 'Tin Symphony' it is a specific 'other' of Australian history, the Irish 'who, more intimate than my intimacy, stirs me', 'who agitates me'.[32] The crowd becomes, with Ned Kelly, the collective carousing underdog. He foot-taps his way into the egalitarian Australian psyche while playing the part of the externalized agent of the frontier. And from the ongoing anxiety Australians have over displays of what defines the national character, a form of tinniness is expressed via Ned's vacant iron-masked stare into the distance and in the corrugated iron skin of the vernacular architecture that so effectively enabled colonial flourishing in an other-worldly land. In the opening 'frontier' of 'Tin Symphony', then, Country is silenced by the din of the jig blaring from the cultural sound system, a jig redolent with the colonial desire for a new but familiar life and a desire for space. Indigenous Country in 'Awaken' becomes land in 'Tin Symphony', a blank canvas for imperial expansion.

III

If 'Tin Symphony' charts a soundtrack of 'patriotism and virtue in turning what was desolate land into an honest and productive industry', on the other side of the frontier Warwick Thornton's *Sweet Country* treats similar history with a deep interrogation of the virtue of the colonial creation myth.[33] The film tells the story of the manhunt and trial following the shooting in self-defence of Harry March, a white property owner, by Sam Kelly, an Aboriginal stockman from a neighbouring farm. In the film, Warwick Thornton spends 108 minutes dwelling on the violence of the frontier that in 'Tin Symphony' has a duration of around eight seconds. Any comparison between the two needs to be framed with this in mind, noting from the start the difference in intent of the productions: the first 'an opportunity for the [Olympic] host [city and nation] to show its best side to the world', and the second to use the genre of the Western 'to tell a story about frontier conflict . . . to highlight and expose settler imaginings about land, law, virtue and vice. It is also able to explore the processes – rather than the myths – of colonisation and effect poignant inversions.'[34]

It is notable that the figure of Ned Kelly plays a role in both productions. In 'Tin Symphony', Ned is the larrikin outlaw dancing across the landscape to enforce a new and singular order over the 250 or more Indigenous Countries each with their lore and some of which are more than 50,000 years old. In *Sweet Country*, towards the climax of the film, an outdoor cinema provides entertainment to the townsfolk whose enjoyment is contrasted with the exhausted anger of local police sergeant, Sergeant Fletcher, who has been outwitted by the fugitive Sam Kelly. The entertainment is an early Ned Kelly gang silent film with the projectionist announcing to the cheers of the crowd that the Kelly Gang 'escapes the long arm of the law'.[35] Sergeant Fletcher stops the show and angrily sends everyone home. The film's action ceases and, projected on the screen, we see the acetate film stock begin to melt. Consequently, there is no capture and hanging of Ned Kelly to come. The Kelly Gang escape is suspended. In *Sweet Country* it is not long, however, before Sam Kelly is compelled to turn himself in and be tried for shooting Harry March.

In contrast to 'Tin Symphony' where Ned Kelly and corrugated iron are activated by an Irish jig, in *Sweet Country* there is no film music except for the closing titles. Rather, it is incessant wind over an arid outback Australian landscape that builds the sonic world. As Warwick Thornton has noted, however: 'The truth is there is music in the film. What it is . . . it's the atmosphere of the land. You know what I mean? The way that the wind blows through the pine trees. It's a note; it's a beautiful sound.'[36] *Sweet Country* is noisy, as if through the wind Country insists on having a voice: as wind, spirit and breath. In locations free of settlers' structures, the wind activates grasses and trees, while around settlers' homesteads and the town, the wind activates tin roofs and walls and water tanks in a continuing, clanging, clunking, creaking, unsettling clatter.

Typically, this clatter is understated rather than overt, producing a subliminal uneasiness so that neither audience nor characters are easy in their skin. Perhaps the homeliest of scenes in the film are those where the fugitive Sam Kelly and his wife Lizzie are on the run and living from the land. The wind blows and the notes are soft through grasses and the wispy desert pines. In town uneasiness is pervasive and for Sam Kelly, after turning himself in and awaiting trial in the corrugated iron lock-up, his leg jitters nervously in response to the wind rattling the tin as well as the rhythmic blows of a hammer that constructs a gallows even before the court is convened. In 'Tin Symphony', Ned Kelly danced a new law across the land. That same frontier energy is present in *Sweet Country*, but the uneven beat of tin against the even hammer blows draws the audience together in foreboding. There is an altogether different unison to that of the hand-clapping crowd of the Opening Ceremony, as we (writing from a dominant white, settler perspective) bear witness to our forebears, the loosely arranged court-scene gallery, that is depicted in the film. And if we identified with the underdog in Ned Kelly covered in his black armour, this is far from the case with Sam Kelly in his black skin.

The tin-clanging uneasiness in *Sweet Country* is typically from non-diegetic sound. An exception comes in the outdoor court scenes that feature the witness seat. The unevenness of power relations before the law is particularly acute when Lizzie and then later Sam Kelly himself are called to bear witness. As Lizzie reluctantly takes the seat before the cultivated and out-of-place Judge Taylor, the wind becomes diegetic. While the clang of tin continues off-screen, onscreen it flaps the pages of the judge's notebook, and when the judge asserts that Lizzie must answer a question, the response comes in the intensified sound of the wind, Lizzie's nervous breathing, and the knocking of the door of a corrugated iron shed in the background. To the exasperation of Judge Taylor, only the wind responds. When Sam Kelly takes the stand (see Plate 5), his chains clink, dogs bark in the echoing distance, the metallic buzz of cicadas starts up, and in the background of the frame the same door makes the tin shed resonate. All the while tin clangs off-screen. There is a sense of anxiety in the breath of the land, obstructed by tin structures which it rattles, and in the short term cannot challenge. Ultimately, Sam Kelly is found not guilty of murdering Harry March and is freed because he acted in self-defence. There is a strong resonance with the interrupted Kelly Gang film with its unplayed final scenes. The townsfolk murmur in dissent. With the gallows denied, Sam Kelly, riding in a sulky with Fred Smith, Lizzie and his niece Lucy, is escorted out of town by Sargeant Fletcher. After a short distance Fletcher decrees, 'You'll be right from now on.' A storm gathers. The air sizzles. Sam slumps back, hit by a sniper's bullet.

The wind roars. A dog barks.

In *Sweet Country*, the wind is the voice of Country which sonically activates the land through which it flows. In town and around homesteads, the ubiquitous corrugated iron provides the material for activation. Just as the characters, whether settlers or Aboriginal, appear uncomfortable in their skin in these environments, so the arhythmic rattle of tin forms a second skin that makes extimate in the audience the inner anxieties of the situation. Often the rattling tin sits in the sonic background incessantly annoying, though not necessarily consciously so. It is a relief when out of settlements the wind works softly on long grass. For an Australian audience, tin can similarly go unnoticed in the visual realm. Tin is an obvious choice for creating the set depicting a town and homesteads in 1930s central Australia. In its visual form, it fits in easily; it is comforting in that sense. Only when it comes loose and is allowed to be activated by the wind does it in turn unsettle the audience who is already in a state of foreboding when placed in the familiarly unfolding scenario of frontier violence.

IV

The two case studies of the sonic activation of tin reveal the affective polyvalence of the material's aural resonance. In 'Tin Symphony', tin is activated by a folk-rock Irish jig which joins the Sydney Olympics Opening Ceremony crowd together in a lively, hand-clapping feeling of unfolding possibility. The gun-blazing gang of Ned Kellys clear the land, with a short excerpt of Aboriginal song at the beginning of the segment that is snapped off by a whipcrack, the only sign of resistance. From then on, the 'Tin Symphony' is, as the etymology of the word suggests, a unison of sounds that unites an audience in enthusiastic national feeling. In *Sweet Country*, however, while tin, in the form of corrugated iron, is ubiquitous and clearly important in establishing shelter for 'settlers' on the frontier, its activation by wind does not unify in a positive, patriotic mode. Instead, it unifies the audience and the film's characters in an uneasy, clanging, rattling assemblage. Indeed, through feeling unsettled, the rattling of loose tin in the wind tends to gather loosely. The audience in its alienation is affectively separate, with a sense of belonging with the land disrupted and the virtue of the nation undone. Depending on the subjectivity of the person in question, this separation has differing affects and effects. For Aboriginal characters, *Sweet Country* depicts the separation of communities from their Country and lore, while for the invading settlers the separation is between their culture and law and the country they desire. Tin, activated by wind as spirit of the land, gives an affective voice to this spectrum of effects.

Thinking affect through sound brings into relationship sounding materials, their activators, subjectivities, as well as a host of actors that already have been giving form to historical conditions in place. This is a complex and rich mix that provides a path into making some sense of our perception and reception of cultural products. The concept of extimacy, meanwhile, provides a process model for considering the embodiment of affect: a topological surface or skin that is at once a relay between psyche and sociopolitical milieu, disrupting the separation of public and private, yet no less mysterious for expressing such a process in words; something better felt as well as reasoned. Such a feeling is somewhat like tin, tinny, in the Australian landscape: intimate and comfortably familiar yet rattling the certainty of comfortable knowing.

Notes

1 E. S. C. Weiner and J. A. Simpson, *The Oxford English Dictionary*, 2nd edn (Oxford, UK: Clarendon Press, 1989).

2 Weiner and Simpson, *Oxford English Dictionary*, 116 [Def. 4b].

3 Weiner and Simpson, *Oxford English Dictionary*, 118 [Def. 2a].

4 Weiner and Simpson, *Oxford English Dictionary*, 116 [Def. 5].

5 Michel Chion, *Audio-Vision: Sound on Screen*, trans. Claudia Gorbman and with a foreward by Walter Murch (1990; New York: Columbia University Press, 1994), 5.

6 Weiner and Simpson, *The Oxford English Dictionary*, 115 [Def. 2a].

7 Philip Goad, *Troppo* (Balmain: Pesaro Publishing, 1999), 25.

8 Adam Mornement and Simon Holloway, *Corrugated Iron* (London: Frances Lincoln, 2007), 157.

9 Philip Drew, *Leaves of Iron* (Sydney: Law Book Company, 1985), 76.

10 Mornement and Holloway, *Corrugated Iron*, 10.

11 Peter Myers, 'Corrugated Galvanised Iron', *Transition* 1–2 (1981): 24.

12 Myers, 'Corrugated Galvanised Iron', 25.

13 Mornement and Holloway, *Corrugated Iron*, 159.

14 Jacques Lacan, *The Ethics of Psychoanalysis* (1986; London: Routledge, 2008), 171.

15 D. H. Lawrence, *Kangaroo* (1922; London: Heinemann, 1970), 349.

16 1984dreaming, 'James Blundell – Rain on a Tin Roof' [music video], You-Tube (2014), https://youtu.be/ETPuJpfi5Pc (accessed 7 December 2021).

17 Glen Phillips, 'February', *Journal of Australian Studies* 28, no. 83 (2004): 130, copyright © International Australian Studies Association reprinted by permission of Taylor & Francis Ltd, http://www.tandfonline.com on behalf of International Australian Studies Association.

18 Kim Scott, *Taboo* (Sydney: Pan Macmillan, 2017), 48.

19 Jane Bennett, *Vibrant Matter* (Durham, NC: Duke University Press, 2010), 100–1. On the term galvanizing in this process, Nicholas Taylor, 'Already Monstrous' (PhD Diss., Southern Cross University, Australia, 2020), 21–2.

20 Jacques-Alain Miller, 'Extimité', in *Lacanian Theory of Discourse*, ed. Mark Bracher, Marshall Alcorn, Ronald Corthell, and Françoise Massardier-Ken-ney (New York: New York University Press, 1994), 75.

21 Tony Hughes-d'Aeth, 'Kim Scott's *Taboo* and the Extimacy of Massacre', *Journal of Australian Studies* 45, no. 2 (2021): 2.

22 International Olympic Committee, 'Sydney 2000 Opening Ceremony – Full Length' [film], YouTube (2020), https://youtu.be/qsLLzL27hYA (accessed 7 December 2021); Mornement and Holloway, *Corrugated Iron*, 159.

23 Country, with a capital C, is an Australian First Nations concept that 'includes humans, more-than-humans and all that is tangible and non-tan-gible and which become together in an active, sentient, mutually caring and multidirectional manner in, with and as place/space'. See Bawaka Country and others, 'Co-becoming Bawaka', *Progress in Human Geography* 40, no. 4 (2016): 456.

24 Jackie Hogan, 'Staging The Nation', *Journal of Sport and Social Issues* 27, no. 2 (2003): 100–23.

25 Damien Halloran and Maria Millward, 'Under Southern Skies', in West, *Opening Ceremony*, 37.

26 Ian Cooper, 'Tin Symphony (Opening Ceremony) (Games of the XXVII Olympiad, 2000 Sydney Olympics) (Extended Version)' [music video], You-Tube (2015), https://youtu.be/bTiUW7x0iMc (accessed 7 December 2021).

27 David West, *Opening Ceremony of the Games of the XXVII Olympiad in Sydney 15 September 2000 – Media Guide* (Lausanne: International Olympic Committee, 2000), 33.

28 Sidney Nolan, 'Ned Kelly' [enamel paint on composition board], Canberra: National Gallery of Australia, 1946, https://searchthecollection.nga.gov.au/object?uniqueId=28926 (accessed 7 December 2021).

29 Nigel Jamieson, 'Make It Relevant', *Australasian Drama Studies* 54 (2009): 113.

30 West, *Opening Ceremony*, 33.

31 Rob Garbutt, *The Locals* (Bern: Peter Lang, 2011), 92–3.

32 Miller, 'Extimité', 77; Jacques Lacan, *Écrits* (1977; London: Tavistock, 1982), 172.

33 Annemarie McLaren, 'A Many-Sided Frontier', *Australian Historical Studies* 50, no. 2 (2019): 242.

34 Mornement and Holloway, *Corrugated Iron*, 159; McLaren, 'A Many-Sided Frontier', 244.

35 Charles Tait, *The Story of the Kelly Gang* (1906) [film], Canberra: National Film and Sound Archive (2021), https://aso.gov.au/titles/features/story-kelly-gang/ (accessed 7 December 2021).

36 Dylan River and Warwick Thornton, 'Sweet Country (Special Features: Dylan River and Warwick Thornton)' (2018) [interview], Australia: Universal Sony Pictures.

Entry 10

Vroom

Andrija Filipović

According to the website writtensound.com, automotive onomatopoeia includes YEEeeEEeeEEeeEEeeEEee! (sound of a loose belt in a car engine), woh woh woh woh (sound of a bad bearing in the timing belt tensioner), vooRRRR, vooRRR, vooRRR (engine revving up and down), varoom (sound of a fast driving car), to name just a few. I take *vroom*, which according to the same webpage is simply the sound of a car, to signify all of these sounds. In this entry, I analyse the sound of cars and traffic, of the urban soundscape, in the context of the postsocialist condition in Serbia and, particularly, in Belgrade. I do so ontosonographically, a conceptually led methodology which explores the ways in which beings and assemblages as multiplicities of interrelated beings and forces affectively *become* in and through sound.[1] Postsocialist infrastructure that enables and disables automobility is of deciding importance for the constitution of the urban soundscape, so much so that, from the point of view of sound intensity, the city's soundscape, automobility and infrastructure produce a single plane of immanence. One can analyse the ways of *being* and *becoming* in the city through the changes in the intensity and prevalence of certain sounds, in the case of this entry the sounds of privately owned vehicles and public transport, the sounds I am collectively calling *vroom*. Postsocialist infrastructure and urban soundscape are inextricably interrelated, constituting the hypercomplex assemblage that we call a city. Ontosonography is a methodology that maps intensities of affective forms of *being* and *becoming*. In the case of the city of Belgrade, intensities of affective forms of *being* and *becoming* are co-constituted by a historical experience of self-governing socialism and transition towards the free market liberally democratic society as found in the West. These affective forms are decidedly defined by the postsocialist condition. *Vroom*, then, is the sound affective form of the postsocialist condition in Belgrade.

Immanence, affect, ontosonography

Vroom is understood through the new materialist ontology that foregrounds the concepts of immanence and affect. Immanence is defined as a single plane of matter on which the cultural and natural, semiotic and material, discursive and corporeal *become* at the same time, and are not only interdependent but of one and the same matter. Signifying and non-signifying matter is of one kind, it is one matter, and when this is taken to its logical conclusion *becoming* is defined by a differentiation in degrees and not in species. This concept of immanence thus disturbs any kind of substantializing binary opposition, while difference as such *becomes* ontological un/ground. It is an ontology that is described in the formula pluralism = monism.[2] Multimaterial bodies on the plane of immanence are in the incessant process of *becoming* through acting and being acted upon, which is a way to say that bodies are constituted through affect. Affect is partly the capability of a body to act and to be acted upon,[3] and is, in addition, the potentiality of a body – the pure and intensive virtuality that is immanent to a body.[4] As such, affect is pure relationality. Considering that every body of any materiality whatsoever is equally relational as any other, it follows that both cultural-discursive and natural-material aspects of a single matter are equally affective. In other words, what enables conceptualization of immanence as such is the affect as pure relationality.

Following these insights, I argue that both noise (as non-organized sound) and music/speech (as organized sound) are equally (non)meaningful and equally yield to analysis. Namely, if any kind of binary opposition falls apart from the point of view of affective immanence, so does the opposition between non-organized and organized sound, meaningless and meaningful sound, noise and music/speech. Instead, there is only an immanent soundscape that is constituted through affect. Soundscape is formed as both actual and virtual relationality, that is as both actual and virtual affect. More precisely, one can analyse different layers of soundscape materiality rather than kinds or species of sounds, which is a view that leads to substantializing binary oppositions and the metaphysics of identity. These acoustic layers are at the same time signifying and non-signifying, and they mutually and affectively condition each other through being and becoming in inter-relation. Meaningless noise and meaningful music/speech are only some layers among the multiplicity of strata on the plane of affective, that is, actual-virtual relational soundscape. As I have claimed elsewhere, the noise produced in the urban soundscape is as signifying as music/speech, or, music/speech is as non-signifying as noise is usually perceived.[5] The noise of urban soundscape is equally signifying and non-signifying, and it acts and is acted upon by the multiplicity of the bodies that move through the urban soundscape. Non-acoustic and acoustic bodies constitute each other leading each other into the processes of *becoming* by affectively relating on multiple levels of materiality.

The ways in which the sounds of cars and traffic, in general, form the urban soundscape of postsocialist Belgrade are mapped through the concept of *ontosonography*. The concept of ontosonography names the forms of *being* and *becoming* through and in sound, and it will be used to map the affective immanent soundscape of (non)life under the postsocialist condition, showing the ways in which cars, humans, and other beings interrelate to form an ecology of the urban soundscape. Thus, ontosonography is not ontology of sound but ontology *through* sound. Through analogy with the sonar and medical ultrasound, ontosonography probes the immanence that *beings-in-becoming* affectively create and maps the intensities depending on how deep the sound can spread and on the types of materialities it meets on its way. Ontosonography offers insight into the multiplicity of interrelated rhythms of *becoming* and maps the diffusion of intensities of these rhythms in the strata of the plane of immanence that is the urban soundscape. The most ontosonographically intensive part of Belgrade's soundscape is the noise made by the traffic, what I call *vroom*, which means that cars and other means of automobility dictate the affective becoming of (non)human forms of life in the urban sound environment. Ontosonography also maps the state of infrastructure, exposing the most intensively used and deteriorated parts of urban infrastructure. Ontosonography reveals the ways in which infrastructure conditions what is recognized as the most dominant form of life by the virtue of being the loudest in the urban soundscape and the most mobile in the urban landscape. Such a form of life turns out to be the postsocialist consumerist one rooted in the exploitation of fossil fuels. It is also a form of life that aims to become other-than-postsocialist by implementation of policies in the process of accession to the European Union, which also leaves audible traces that yield to ontosonographic mapping.

Infrastructure, automobility, vroom

The term 'postsocialist condition' marks the end of the actual experience of a self-governing socialism in Serbia and names the present experience of economic, political, social and cultural transition towards what is defined as contemporary European state and society. The end of a self-governing socialism was marked by the civil wars in Slovenia, Croatia and Bosnia and Herzegovina in the 1990s that broke apart the Socialist Federal Republic of Yugoslavia into six constitutive republics. In Serbia, the dissolution of Yugoslavia was reflected in international sanctions and an autocratic regime of Slobodan Milošević until 1999, when the Kosovo war broke out and NATO carried out a bombing campaign to end it. With the end of Milošević's dictatorship in 2000, Serbia began reforms in order to join the European Union, opening negotiation chapters depending on EU reports on reform development. These reforms, most importantly, include economic reforms towards the liberalization of the market, with privatization of what was once owned by the state and the workers as the backbone of these changes. What is still being privatized are mostly large corporations once owned by the workers themselves, an exception being the largest communication corporation (Telekom), the only electric power corporation (Elektroprivreda Srbije) and the public radio-television service (Radio-Televizija Srbije). Liberalization of the market, together with the insistence on private ownership, also entailed the opening of the Serbian economy for foreign investments and imports. The largest economic partner of Serbia is the European Union, hence most of the imported goods come from the EU countries.

Of particular importance for the intensity of *vroom* is the import of cars and other vehicles. Also, one needs to keep in mind that Serbia produces its own brands of automotive vehicles, including cars, buses and agricultural vehicles. The most famous car brand is Jugo, made infamous as 'the worst car ever' during the time of socialist Yugoslavia due to its cheap production and inefficiency. The period of socialist Yugoslavia was also the time when the first issues with automobility were noticed, which became more severe with the progression of decades. The Socialist Federal Republic of Yugoslavia underwent a period of strong economic development in the 1960s and 1970s, and the decades were marked by an increase in the number of personal cars across Yugoslav republics. Maja Fowkes quotes a letter from the 1970s to the editors of a Croatian architecture journal that reads: 'A wave of motorization has gripped our country. Ten years ago it was possible to count on your fingers the number of personal cars, except official ones. Today our streets are crowded with cars and every year there is less space.'[6] Environmental issues caused by the increase of the number of cars were noted in neo-avant-garde art of the time, in the works by TOK, a group of conceptual artists from Zagreb, Croatia, who left 'traces of car tires along the lines of parked cars, and in this way marking the space on the pavement left for pedestrians.'[7] There were ecologically conscious neo-avant-garde groups in Belgrade – Clear Streams Family – and in Ljubljana – OHO. Cars were seen as 'polluters of air and noise' and 'unjustified means of transportation', which also caused 'compulsion to destroy landscapes in order to build more and wider roads', as well as 'unheard-of traffic jams, the uneconomic use of fossil fuels'.[8] While

TOK was active in Zagreb and pointed out issues with automobility in that city in their artworks, the same situation at the same time was to be found in Belgrade too, which was seen as the capital of the whole of Yugoslavia. The city itself underwent a project of modernization in the decades of socialist Yugoslavia, with the largest municipality of Novi Beograd built in the area that used to be a river swamp.[9] Novi Beograd is now the most numerous municipality in both residents as well as automobiles.

Infrastructure for automobility built in these times has not been expanded in any substantial way until the present day. What was built then rarely remained in original condition. It mostly deteriorated because of the lack of maintenance due to, first, the absence of interest (since it was working properly), then to the international sanctions (during the dissolution of Yugoslavia and the Kosovo war under Milošević in the 1990s), and finally, to the transitional state and the lack of investment interest from 2000 until today. Or, rather, it was not the lack of investment interest that influenced the deteriorating state of infrastructure, but the push towards the privatization of once common and public property. The idea of the public good was no longer seen to be viable and as a result there was (and is) less money for investment in public infrastructure. The deteriorating state of public infrastructure is part and parcel of the postsocialist condition in Belgrade and Serbia, and it is a clear sign of movement towards a liberal capitalist market which pivots on the difference between private and public, with private property being the only one worth investing in. The streets that both enable and disable movement of cars and other forms of transportation, and thus produce and shape the thick affective fabric of the urban soundscape and with it the *vroom*, are decidedly marked by the transitional state of postsocialism.[10]

The streets reveal the postsocialist condition by their very affective-relational materiality, either through their deteriorated state or by being built to withstand lower numbers of vehicles, which results in traffic jams and accidents. The environment around the streets is also important for the constitution of the urban soundscape and its most affectively intensive aspect, that is the *vroom* of traffic noise. The second decade of the twenty-first century has been marked by so-called investment urbanism, which is defined as 'the kind of city development planning which puts the investor and his profit interests at the center of decision making. . . . It is really not hard to imagine what the new buildings will mean, besides being an additional load on the already inadequate infrastructure'.[11] It means, besides detrimental ecological effects, the increase of propagation of noise through cut out and other sound effects.[12] Most of the buildings constructed under the banner of investment urbanism are architecturally designed in such a way as to maximally use the allotted space, which means that no space is left for trees or any other objects.

The urban space is reduced to streets and buildings without anything in between to alleviate the intensity of traffic noise – the intensity of *vroom*. On the other hand, there are a large number of buildings in between these new ones constructed during the time of socialist Yugoslavia. These were made quickly and with cheap materials to accommodate the influx of workforce needed for the project of modernizing the city during the 1960s and 1970s. As natural barriers such as trees or parks are destroyed under investment urbanism, traffic noise more easily permeates older buildings, and the sound border between the inside and outside, and public and private, becomes much more affectively porous.[13] *Vroom* creates its own immanence extending across what is usually recognized as inside/private and outside/public.

What is present on the streets of Belgrade is a multiplicity of imported vehicles such as cars and buses, and domestically produced vehicles, including agricultural ones. Both foreign and domestic vehicles are of various age, mostly older models since these were cheaper to import or, if looked at from a local perspective, newer domestically produced models were unaffordable. Another aspect of perpetual socio-economic transition, together with the lack of buying power to replace older vehicles, is the migration to Belgrade.

Belgrade is the centre of economic activity and it numbers almost 1.7 million residents, with almost 600,000 registered cars and over 100,000 other vehicles, according to a report from 2019.[14] The latest partial report for the third quarter of 2020 notices an increase in the numbers of all vehicles except the buses.[15] The result is that the whole of Belgrade is marked by increased traffic and, in several areas, by heavy traffic congestion and consequently heavy noise pollution, the most intensive aspect of *vroom*.

The main reason for the larger number of vehicles lies in the increase of city population and accessibility of cheap imported cars. According to the president of the New Car Import Association, 800,000 old cars were imported since 2014.[16] A state ban on the import of cheaper cars with Euro 3 and Euro 4 motors from the European Union was announced for the purpose of protection of the environment, as these cars 'not only produce dangerous exhaust fumes but are an environmental danger as such', according to the then actual Minister for Environmental Protection.[17] European emission standards for vehicles were introduced in 1992 and the standards regulate emissions of nitrogen oxides, total hydrocarbon, non-methane hydrocarbons, carbon monoxide and particulate matter. Vehicles that do not comply to the standards are forbidden from being sold in the European Union.

The latest stage is Euro 7, which is currently being discussed, to be introduced by 2025. According to available information, it aims to lower production of carbon monoxide from 1 (as in Euro 6d) to 0.1–0.3 grams per kilometre and nitrogen oxide from 0.060 to 0.030 grams per kilometre.

Euro 3 (approved in 2001) and Euro 4 (2006) show significant differences from Euro 5 (2011) and especially from the latest Euro 6d standards (2021). Emissions of carbon monoxide in Euro 3 were 2.3 grams per kilometre, while in Euro 6d is 1, nitrogen oxide 0.5 while in the latter 0.060 grams per kilometre. What this means is that most of the traffic in Belgrade is made by the cars that pollute more and are even noisier than the newer models. And it is not only the cars that are the issue when it comes to the pollution and the production of noise. The public transportation that includes mostly older, domestically produced models of buses is one of the more important contributors to *vroom*.

The intensity of urban noise in Belgrade is measured in spring and autumn cycles by the Secretariat for the Protection of Environment, a city government's body. The Secretariat publishes on its website monthly reports on urban noise and yearly integral reports that include the state of air, water, earth, noise and activities on the protection of the environment in each municipality. Each monthly report on noise includes the satellite image of measurement spots in the city with the results of measurement given in a table overview.

Measurements are taken in each municipality near the largest streets, and tables are given for day, evening and night measurements together with maximal values for each acoustic zone. These maximal values are thresholds for what is perceived as noise in these neighbourhoods given that the urban soundscape is always already present through incessant background hum of variable intensity. Maximal values vary according to acoustic zones in which they are measured, so sound becomes noise only if the high intensity of sound is located near particular objects such as apartment buildings or, better yet, individual homes, hospitals and kindergartens. Thresholds are thus produced through particular techno-scientific technologies for management of both affectivity of urban soundscape and human and non-human beings.

Acoustic zones are chosen as representatives of city zones of different use and along the most important streets. An allowed value for the day and the evening is fifty-five, while for the night is forty-five. The latest report is for October 2020, and it shows that the measured noise on almost every measurement spot exceeds the maximum values for all three periods. Analysing the last three reports from 2016, 2017 and 2018, it can be noted that each report stresses that measured levels of communal noise in the given period were not only high but noticeably above the maximum allowed values. In the report from 2016, levels of noise were above required levels on twenty-eight measurement spots out of thirty-five, the same in 2017, while the report for 2018 claims that there is an excessive noise in a large number of measurements spots during both night and day, and that it is irrelevant whether it was a spring or autumn measurement cycle.[18] One can conclude from the reports that intensive traffic noise – *vroom* – is a permanent feature of the affective fabric of the urban soundscape and, in actuality, its most intensive aspect, and that particular techno-scientific management technologies have been developed in dealing with it, such as traffic control through various means (circular cross-ways, bypasses) and acoustic zoning which entails proscribed and allowed activities.

A report from 2018 also notes the percentage of the urban population endangered by the traffic noise. Numbers go from 11 per cent near Perside Milenković Street in the Senjak municipality, which is considered an elite residential part of the city, to 55 per cent during the day in several places and municipalities. The report concludes that the percentage is very high and that certain measures for alleviation need to be taken. All three reports agree that the source of noise needs to be controlled. Since the main noise polluter is the traffic, control can be performed by 'automatic regulation of traffic and synchronization of traffic lights, replacement of old cross-ways with circular ones, construction of bypasses, displacement of the last and the first bus stops, increase of number of zones with traffic limits, maintenance of roads, etc'.[19] The report also advises to follow European automotive standards regarding the motor noise and that noisier cars need to be excluded from the traffic. Considering the number of imported Euro 3 and Euro 4 cars, such alleviation measures are obviously far from being implemented.

One of the advised control measures is the acoustic zoning of the city territory, as it 'enables of selection of permits for performing certain activities in specific city zones'.[20] According to Beoinfo news from August 2020, Belgrade will set acoustic zones for all seventeen of its municipalities by January 2021, where each will have a maximum allowed level of noise depending on whether there is a hospital, daycare centre, school or some other institution in the vicinity.[21] From the point of view of ontosonography, acoustic zoning has the effect of shaping the forms of life through the sounds these forms produce, as well as of shaping the sounds that affect the *becoming* of forms of life. As already noted, investment urbanism together with the state of infrastructure and individual vehicles defines the affective fabric of urban soundscape. However, as can be gleaned from the reports of the Secretariat for the Protection of Environment, the quietest parts of the city are the most affluent ones.

Perhaps the most direct relation between the affectivity of *vroom*, that is the urban traffic noise, and class is to be found in the case of Old Sava Bridge built in 1942 during the German occupation, which is now located at the very heart of the Belgrade Waterfront project, an epitome of investment urbanism. Belgrade Waterfront 'represents a combination of commercial and residential luxury space. It is a foreign investment project for which the state gave the land and offered clear support. If realized in the planned scale it will significantly transform that part of the city through the process of profitable gentrification. The space will be dedicated to the members of elite and foreign citizens.[22] The project has been controversial from the very beginning for various reasons, including the influx of big foreign capital that completely changed the neighbourhood landscape and old local ways of life.[23] The city government has made plans to remove the bridge and relocate it to the nearby park. In other words, those who manage the postsocialist transitional condition better economically have higher chances of living in acoustic zones that are less defined by *vroom*, traffic noise, and more by the sounds of other human beings and animals (coded as 'natural' and thus more appealing), and sometimes car alarms, which signify and signalize private ownership in the affective fabric of the urban soundscape. 'Natural' forms of life can form their own affective existential refrains without much intrusion from the outside, or at least such an intrusion can be tolerated for longer periods of time. Those living in buildings constructed in noisier acoustic zones that are also in less affluent municipalities, and those living in the buildings constructed in the times of investment urbanism, have lower chances of mollifying the affects of the urban soundscape. They are more exposed to the affective immanence of the urban soundscape, to the affective force of *vroom*, for the better or the worse. They are more exposed to the multimateriality and pluritemporality of assemblage of forces that is the city as such.

Conclusion

Ontosonographic analysis of the postsocialist urban soundscape reveals defining spatial and temporal features of transitional society, and the ways in which such an assemblage is *becoming* through an intertwining of the multiplicity of affective forces. The oftentimes overwhelming traffic noise, the above-threshold intensity of *vroom*, points to several aspects of the postsocialist condition and the urban infrastructural and soundscape assemblage. It shows the presence of various temporalities in automobility ranging from contemporaneity in the form of electric cars and buses to the socialist past in the form of older models such as Jugo. It also complicates the simple linear temporality given the presence of imports from the EU of older models, which are newer than the ones already present on Serbian streets. To this, one should also add the promise of an EU future materializing through imported goods and undertaken or desired reforms. Such a future is as noisy as the oldest models or as silent as the newest electric vehicles. The urban soundscape of Belgrade is a non-linear temporal assemblage where past, present and future intersect to constitute an aspect of hypercomplex relationality of strata on the plane of immanence.

As far as space is concerned, the ontosonography of *vroom* reveals the multimateriality of the bodies that move through it and define it. One important aspect is the incessant insistence on private ownership. As already noted, the ownership of private vehicles has been a sign of personal prosperity since at least the 1970s, but today it carries additional layers of meaning. Owning a vehicle must be understood in relation to the whole urban assemblage, which includes the state of the infrastructure – the crumbling roads, the lack of parking space, the quality of air, the relation to the fossil fuels industry local and global.

Space, and the bodies in it, turn out to be local, regional, and planetary at the same time, on the one hand. On the other hand, privatization is what defines the postsocialist condition, and it is an aspect which most intensively participates in the urban soundscape and urban ecology. Bodies and the sounds they produce are sequestered in their own spaces and the spaces that are privately owned. Moreover, entire urban zones will be sequestered according to the intensity of the urban soundscape and proscribed thresholds, the intensity of *vroom*, following the private-public dividing line, and only for the acoustic good of some.

Notes

1 See Gilles Deleuze, and Félix Guattari, *A Thousand Plateaus: Capitalism and Schizophrenia*, trans. Brian Massumi (1980; Minneapolis and London: University of Minnesota Press, 1987), 232–309.
2 Deleuze and Guattari, *A Thousand Plateaus*, 20.
3 Deleuze and Guattari, *A Thousand Plateaus*, 261.
4 See Brian Massumi, *Parables for the Virtual: Movement, Affect, Sensation* (Durham, NC and London: Duke University Press, 2002).
5 See Andrija Filipović, 'Resonant Masculinities: Affective Co-production of Sound, Space, and Gender in the Everyday Life of Belgrade, Serbia', *NORMA: International Journal for Masculinity Studies* 13, nos 3–4 (2019): 215.
6 Maja Fowkes, *The Green Bloc: Neo-avant-garde Art and Ecology under Socialism* (Budapest: Central European University Press, 2015), 136–7.
7 Fowkes, *The Green Bloc*, 136.
8 Fowkes, *The Green Bloc*, 137.
9 See Brigitte Le Normand, *Designing Tito's Capital: Urban Planning, Modernism, and Socialism in Belgrade* (Pittsburgh: University of Pittsburgh Press, 2014).

10   For 'temporal fragility' of infrastructure, see Kavita Ramakrishnan, Kathleen O'Reilly and Jessica Budds, 'The Temporal Fragility of Infrastructure: Theorizing Decay, Maintenance, and Repair', *EPE: Nature and Space* 14, no 3 (2020): 674–95.

11   Sandra Petrušić, 'Temeljna betonizacija – Naprednjačka humanizacija grada', 2018, https://nedavimobeograd.rs/nin-temeljnabetonizacija-naprednjacka-humanizacija-grada (accessed 8 February 2021); see also Jelisaveta Petrović and Vera Backović, eds, *Postsocialist Capitalism: Urban Changes and Challenges in Serbia* (Belgrade: Institute for Sociological Research, 2019).

12   Jean-François Augoyard and Henry Torgue, *Sonic Experience: A Guide to Everyday Sounds* (Montreal: McGill-Queen's University, 2005), 30.

13   See Filipović, 'Resonant Masculinities', 219–22.

14   Dušan Gavrilović, *Opštine i regioni u Republici Srbiji* (Beograd: Republički zavod za statistiku, 2020), 297.

15   Republički zavod za statistiku, 'Prvi put registrovana drumska motorna i priključna vozila i saobraćajne nezgode na putevima, III kvartal 2020', 2020, https://www.stat.gov.rs/sr-latn/vesti/20201123-prvi-put-registrovana -drumska-motorna-i-prikljucna-vozila-i- saobracajne-nezgode-na-putevim a-iii-kvartal-2020/?s=1502 (accessed 8 February 2021).

16   Nezavisne, 'Zabrana uvoza Euro3 vozila u Srbiji realna do 2021', 2019, https://www.nezavisne.com/automobili/auto-novosti/Zabrana-uvoza-Euro3 -vozila-u-Srbiji- realna-od-2021/568598 (accessed 8 February 2021).

17   Nezavisne, 'Zabrana uvoza'.

18   Nataša Petrušić, *Kvalitet životne sredine u Beogradu u 2018. godini* (Beograd: Sekretarijat za zaštitu životne sredine, 2018), 153.

19   Petrušić, *Kvalitet životne sredine*, 156.

20   Petrušić, *Kvalitet životne sredine*, 157.

21   Beoinfo, 'Beograd propisuje akustičke zone i dozvoljen nivo buke u njima', 2020, http://rs.n1info.com/Vesti/a630869/Beograd-propisuje-akusticke -zone-i-dozvoljen-nivo- buke-u-njima.html (accessed 8 February 2021).

22   Vera Backović, *Džentrifikacija kao socio-prostorni fenomen savremenog grada: Sociološka analiza koncepta* (Beograd: Filozofski fakultet, Univerzitet u Beogradu, 2015), 183.

23   See Herbert Wright, 'Belgrade Waterfront: An Unlikely Place for Gulf Petrodollars to Settle', 2015, https://www.theguardian.com/cities/2015/dec /10/belgrade-waterfront-gulf-petrodollars-exclusive-waterside-development (accessed 8 February 2021).

Entry 11

Thump

Andrea Avidad

How MUCH my life has changed, and yet how unchanged it has remained at bottom!

Franz Kafka

In the first seconds of Argentine auteur Lucrecia Martel's feature *La Mujer Sin Cabeza* (*The Headless Woman*, 2008), one hears: *footfall, footfall, pant, pant, bark, bark*. A tracking shot frames the lower half of a human body running alongside an unleashed dog whose canine body, unlike the human one, is fully visible on the screen. As the camera brings subjects and environment into visibility, the lower-class cosmos of provincial Argentina takes up the screen: *mestizo* (mixed-race) boys jollily scamper in all directions on an empty canal bordering a dirt road, an unpaved path where neglected public buses sporadically circulate with their dilapidated chassis. The dark-skinned teens wear simple clothes; the barking mammal is no *perro de raza* (purebred dog). The rocky flesh of the road, surrounded by the vestiges of dried out vegetation, saturates the cinematic frame with naked roughness. A cut conjures up the other side of the Argentine class and ethnic architecture: a world of appearances and gazes; a world of fake eyelashes and peroxide-blonde hair; a cosmos of sedans and SUVs. In the universe of the bourgeoisie, light-skinned women speak of aquatic turtles, asserting that the shelled reptiles are nothing but vessels of disease that will destroy the risqué fantasy of a soon-to-be-inaugurated private swimming pool, a *piscina* destined to be disgraced given its 'lamentable' physical proximity to a veterinarian clinic. In the vain eye of the upper middle class, the figure of the 'animal' is drenched with impurity and abjection.

These racialized, class-determined planes of existence, the white bourgeoisie and the *mestizo* working class, seem to collapse into one another as Vero (María Onetto), cocooned in her car, runs over a body (somebody?) (something?) – a body that is never clearly seen – on the dirt road. But what isn't fully seen is heard: *THUMP! THUMP!* Loud thumping sounds etch the anonymous body (*Mestizo* body? Canine body? Perhaps bodies?)

within the diegesis. Notwithstanding their violent appearance, such sounds fail to disclose the ontological specificity of the body or bodies involved in the accident. A blind (or is it mute?) spot; translated into visual terms, the thumps appear as a nebula, as a cloudy silhouette that certainly suggests, and yet, also decidedly refuses to tell.

A brief gesture to find out the identity of the crashed entity (entities?) is swiftly aborted. Vero instead opts for putting her dark sunglasses back on and drives away. From this point onward, Martel's film turns into a labyrinthine web of intimations: what could have happened but didn't; what could have happened and did happen but might not be all that happened; what could have happened but must not be spoken-of. These (im-)possibilities compose the flesh of the film, a film that, staging an unorthodox non-linear time in which the past does not simply disappear but rather remains virtually present in the present, re-enacts, time and again, what remains an opaque occurrence (did Vero leave a boy to die? A mongrel? Both?). What caused the thumps on the road? *Shush!*

We soon realize that to look for veracity would be otiose. Rather than offering a calculus of truth, the film's own idiosyncratic method consists in centring its telling around the crisis of signification inaugurated by the thumping sounds – a crisis gyrating around the unknowability of a disturbance that itself brings about further crises: affective and ethical. After listening to the thumping sounds whose 'source' remains concealed as an eerie absence-presence-absence, in a perpetual movement of displaced reference, Vero becomes essentially de-animated, nearly speechless, at the limits of expression. She becomes a dolled-up automaton whose perfunctory utterances cannot even keep up with the mundane routines and aimless talk of everyday life. She enters a mode of dull affect and de-animated thought. The sonic-affective power of the thumps resonates in Vero's body and head, leaving her in an affective condition of 'headless-ness'. After listening to and feeling the thumping sounds, Vero *is* the headless woman.

These thumps, then, are not mundane thumps. These violent thumps, whose affective resonance de-animates Vero, and throw her into an embodied mood of dullness and headless-ness, must be listened to, must be thought of, as *sound affect*. In my reading of the 'affective thump', a sound affect can be thought of as a sound whose resonance impacts the body and the consciousness of the listener. To my way of thinking, a sound affect is a sound whose affective force infiltrates the knots between cognition, perception and embodied feeling, opening up all sorts of consequences for she who listens. The affective thump, a sound affect, makes audible questions about normative life and its ethical frontiers: What counts as life 'proper' in the eye, ear and flesh of bourgeois subjectivity?

In what follows, I examine the ethics of the affective thump, zeroing in on the connections between sound and affect, and on the cognitive and ethical problems that spring from the arrival of a sound affect whose force compels thought and feeling, felt thought and thinking feeling, to reflect on the muffled but brutal violence of ordinary life in the Argentine universe. I first look at the sonic form of a thump, resorting to Franz Kafka as my sound theorist. I suggest that the sonic form of a thump is constituted by three essential terms: violence, mediation and dullness. I then track the various consequences of the affective thump in Martel's film, employing the form 'violence-mediation-dullness' as the conceptual constellation that guides my analysis.

Sonic form and sound affect: Violence/ mediation/dullness

What makes a thump a thump? A thump presents to the ear a rather curious phenomenon. A body strikes against another body heavily, and yet, the sound produced by the impact – the thump – has the quality of dullness. The causal event producing the thumping sound, in other words, is characterized by abruptness and violence, but the sonic effect, muffled, is enveloped in dullness. Consider Franz Kafka's description of a thump in his celebrated novella *The Metamorphosis* (*Die Verwandlung*, 1915). Kafka writes that the sound produced by his famous human-turned-bug protagonist Gregor Samsa, as the ex-human falls from bed in the morning of his curious transformation, is not that of a real 'crash' (*Krach*) but that of a 'loud thump' (*lauten Schlag*). Kafka endows Samsa's non-human body with a certain stretchiness: 'Gregor's back was more elastic than he had thought, which explained the not very noticeable muffled sound.'[1] It is the materiality of Samsa's bug-like anatomy, of its rear part specifically, which functions as a mediating force that dulls the sound caused by the impact. Samsa also lands on a carpet, a rug that mediates the collision and absorbs the sonic effect emerging from it. The material composing the body of the object, in this case, the knitted fibres of the rug, in contact with Samsa's non-human fleshiness, muffles the sonic effect.

We can extract key terms defining the 'thump-ness' of the thump: violence, mediation and dullness (the fall, the rug, the muffled sonic effect). Causal event and its sonic effect correlate in a somewhat disparate manner in the sonic form of the thump. Thump: both force and absorption at once. And we can also establish these terms – violence, mediation, dullness – as the essential structure for thinking the sound affect 'thump'.

I now turn to Martel's film, focusing on the violence of the acousmatic situation that determines Vero's mode of listening, on the various techniques of mediation employed so as to dissolve the ethical questions made audible by the thumps and on the de-animated, dull affect that vibrates in Vero's body and (lost) head.

Violence: Acousmatic thumps

Vero drives on the road. In the beginning of the scene, framed in a point-of-view shot, the viewer gets a privileged sight: the arid zig-zagged road appears in all its plenitude. The serpentine canal, waterless, embraces the way. To the right, a road sign warns about the possibility of unexpected entries into the roadway by cattle (after all, this is not Buenos Aires but provincial Salta). We see what Vero sees. Were someone or something, or someone and something, to walk into the road (and frame), the viewer would be able to see it (see Plate 6). But right on cue a cut removes this advantaged view from the spectator. The camera cuts to a static medium close-up shot of Vero's profile. We look at her looking at what we can no longer look at. Her gaze: our blind spot.

Two loud thumping sounds emerge from a violent collision on the road. Vero stays in the car. Quite softly, almost inaudibly, she breathes out. The sound produced by the wetness of her mouth, by her tongue slowly moving inside the moist oral cavity, shyly vibrates against a pop song being played from her car stereo (*'Just a bit lonely, just a bit sad . . .'*).[2]

THUMP! Heard but unseen (what was it?). Acousmatic: heard but unseen (how can one know what it was?). In general, the word 'acousmatic' designates a sound whose source is simply out of visibility.[3] Michel Chion writes: 'just listening, without the images, "acousmatises" all the sounds, if they retain no trace to their initial relation to the image.'[4] In his recent theory of acousmatic sound, however, Brian Kane proposes that while the division between the perceptual registers of sight and hearing contributes to the experience of the acousmatic, such sensorial separation is not what determines this peculiar perceptual mode. For Kane, 'the experience of acousmatic sound is epistemological in character, articulated in terms of knowledge, certainty and uncertainty'.[5]

In Kane's model, the acousmatic situation brings into crisis the identitarian self-sameness between the three constitutive 'parts' of a sound: sonic source, cause and effect. He writes: 'the nearly-*but-not-quite* autonomous auditory effect *necessarily* underdetermines attributions of source and cause.'[6] A sound, when heard acousmatically, is haunted by the shadow of a sonic source and a sonic cause that cannot be fully determined by the listener. Thus, what determines acousmatic listening is not so much the split between two sensorial modes (sight and hearing), but, rather, the degree of *acousmaticity* experienced by a listener.

Following Jacques Derrida's notion of *espacement* – a 'spacing' that interrupts 'every punctual assemblage of the self, every self-homogeneity, self-interiority'[7] – Kane describes acousmaticity as a degree of self-difference between the three components of a sound, that is, sonic source, cause and effect. A sound heard in full acousmaticity inaugurates a crisis of signification, for the listener cannot determine the source or the cause of the sound heard.

How can one think about the disturbing acousmaticity of the thumping sounds caused in the collision on the road? First, one might say that it is the sonic source which undergoes the movement of self-difference. In the case of the road, one can put it as follows: the acousmaticity of the thumps emanates from the endless movement of referral and deferral between two forms of life, human or animal, boy or dog. Like a Kleistian dash, a textual gap whose muteness calls for endless interpretation, the acousmaticity of the thumping sounds refers back to all sorts of probabilities: perhaps, the sonic source is human, or perhaps animal, or perhaps both.

But there is another level to this foundational crisis of signification. The acousmaticity of the thumps also brings about a series of questions that transcend the specificity of the incident, asking: What counts as life 'proper' within the highly stratified and racialized organization of life in the town? What are the (deathly) costs of the living norms and norms of life at work in Vero's world? *THUMP!* The sound and the blow. *THUMP!* The blow, the sound and the bruise. But: what is the nature of this bruise? This bruise is not a visible mark on the body, rather, it is connected to the violent affective state that the sounds empower, an affective state linked to ethical questions.

The acousmaticity of the thumps not only compels Vero to think, over and over, about the threshold between 'human' and 'animal' in the specific instance of the collision. The acousmaticity of the thumping sounds transcends the single event: Vero must confront the (un-)ethical consequences of the normalized division between human and human, between human and 'human-animality' that her social class authorizes in the politics of ordinary life.

What is this 'human-animality' in the specific context of the film? For the bourgeoisie, *mestizo* life is nothing but a reservoir of flesh, energy and pulse whose sole destiny consists in becoming a kind of bio-mechanical labourer. For the white bourgeoisie, *mestizo* existence is pure 'animal' labour: reduced to absolute bodily production, in and by the eye of the bourgeoisie, *mestizo* existence is, *for* the bourgeois subject, 'animal' or bodily power meant to be exploited in everyday life through various practices of labour. For the bourgeois gaze, then, the *mestizo* body might be formally human, but this other(-ed) body is ultimately seen as a mode of 'animal', mechanical existence. This class-determined mode of perception invokes and performs the colonial past, continuing to oppose the inherited normative category of the 'human', that is to say, white, 'civilized' or 'developed', to (colonial) notions of nature, animality, and non-white life, notions that understand the 'less-than-human' as forms of life sentenced to extraction, exploitation and oppression.

Notwithstanding the brutal violence inherent in this class-defined modality of perception and consciousness, Martel shows that, far from being an exception, such a mode of seeing, listening and being is the norm that reigns over the space of life in the town. And the normative acquires a kind of invisibility and inaudibility for those who inherit, embody and reproduce it. Normative violence develops a kind of 'objectivity'. Consider Slavoj Žižek's own definition of 'objective violence' as the type of infrastructural violence that resembles a kind of dark matter, a background that, despite looking like a zero level of non-violence, is the 'stuff' that constitutes our always-violent political systems.[8] Normative, objective violence slyly looks like nothing while certainly being something.

With all of this in mind, I want to suggest the following: the acousmaticity of the thumping sounds inaugurates a kind of feedback loop between the incident of the collision and the normalized violence that grounds everything Vero knows, including *herself*. And this process, this motion, posits critical questions: what are the (un-)ethical thresholds that one might end up transgressing because of one's own cognitive and affective investment in a normative context? Vero's normative context of sociality over-determines the *mestizo* 'servant' class as the class of the labourer 'human-animal', thus training bourgeois subjectivity to perceive *mestizo* life as not 'fully' human.

Vero's quasi-automatic decision to continue driving after the incident, without looking back, throws *mestizo* existence into the flat hole of non-life. Her quasi-automatic action exposes to view, to her own view, that the norms of her social class – norms that shape the knot between perception, affect and action; norms that animate the rules of intersubjective practices in a dynamic relational matrix of sociality – might operate as lethal technologies. In addition to being the technologies by which normalized subjection continues to do its insidious work in the rhythms of everyday life, the norms of the class to which Vero belongs to as a loyal daughter are also the technologies by which voluntary social deafness and blindness ripen into acts of letting die.

The acousmaticity of the thumping sounds, then, makes audible in Vero's ear what can be considered an ethical imperative; this acousmaticity functions as an impersonal call that nonetheless addresses her: what must one do after having had the realization that one might have assaulted the last ethical frontier as a result of one's own habituation to congealed normative views of stratified social life? Vero's acousmatic listening cannot be quieted.

Tiny bursts of vulnerability expose her inability to silence the ethical question echoed by the memory of the thumping sounds. For instance, shortly after her incident on the road, she sees a *mestizo* boy collapse on the ground in a football match. The sight is accompanied by the off-screen barking of a dog. One wonders if this image is a displaced (quasi-)re-enactment of what she saw, without fully seeing, on the road. Perhaps the audiovisual image is a paranoiac projection. Or maybe both. Right after witnessing this rather mundane episode, Vero vanishes into the privacy of the public, taking shelter in an unremarkable public bathroom. There, remarkably, she cries. And it is an anonymous *mestizo* worker that happens to be there who is the person who offers her his anonymous sympathy. The man embraces her; their invisibilities face one another. It is during this minute of intimacy, during this minute intimacy, that Vero is finally able to shed tears. She swiftly recomposes herself, utters a rigid '*Buenas tardes*' (good afternoon), and recovers the tragic performance of her social identity.

Yet, Vero's crumbling performance will continue to be punctured by the acousmaticity of the thumps, by the echoes of a past incident that do not fade away, by the sonic remembrance of a non-transparent incident whose perplexing arrival causes an affect of disorientation, a dull affect that disrupts the synchronicity of the self with its own 'self', as well as the self's symmetry with its normative environment. And it is because Vero's dull-yet-violent affect threatens to disorganize the norms of sociality-as-lived that various attempts to mediate and to nullify her ability to sense the ethical issue inaugurated by the acousmaticity of the thumps will be made.

Mediation: 'Dogs', men, frames

Martel's cinematic architecture foregrounds processes and symptoms, incidents and disturbances, undermining the centrality of causes and origins. In her cinematic style, causes are displaced and misplaced; one can only sense their trace. Her focus is on effects and on effects of effects. I would like to give attention to technologies of mediation in the narrative, in particular, to social techniques that mediate acts of listening to a sound whose source cannot be fully determined, that is, the disturbing thumping sounds.

Halfway through the narrative, in the middle of a popular grocery store, Vero tells her husband: 'I killed someone on the road.' The statement comes out of her mouth with foreign assertiveness. Thus far, Vero's explanation with regard to what happened to her car has simply been: 'I ran over a dog on the road', an utterance wrapped in dissimulated scepticism. *Beep, beep, beep.* Pay for your groceries (and for your guilt?).

Chiaroscuro à la Caravaggio. Two silhouettes enveloped in darkness. A static shot shows Vero and her husband driving on the road, looking for the body or bodies involved in Vero's accident. It is night-time so visibility is compromised. But invisibility is welcomed in this case.

'*It is a dog. There; right there.*' Vero's husband's voice declares that the canine carcass is there, right *there*. But 'there': where? No referent to be seen; no 'evidence' to be witnessed (see Plate 7). Out-of-frame, such a 'there' is both everywhere and nowhere. This is a 'there' whose corrosive generic quality betrays its demonstrative function. Even Vero is unable to find the 'there-ness' of this 'there'. '*I think I killed someone*', she repeats. '*You got scared. You ran over a dog. There; over there . . . it was nothing*', her husband reiterates.

'*There . . . right there. There . . . a dog . . . nothing.*' This act of seeing that belongs to the husband attempts to freeze the movement of acousmaticity of the thumping sounds by referring to a 'presence' that cannot be captured in immediacy. It is the replacement of a displacement with another displaced 'presence'; the replacement of displacement with another displacement that is said to not be one. Should one accept this prosthetic act of seeing, this act of seeing and of saying that seeks to mediate one's own acts of seeing and of listening? Should one subscribe to the approach proposed by the husband, an approach that reduces the complex ethical and affective issue initiated by the acousmaticity of the thumping sounds to a matter of (failed) empirical perception? No. The thumping sounds make audible something that transcends empirical optics. But auxiliary acts of listening will attempt to mediate Vero's act of listening. Echoing the words of Vero's husband, her cousin admits to having heard the thumping sounds *many times* on the road. The cousin also admits that the sounds are appalling, a 'horrific' noise that he himself knows very well. But he affirms that 'nothing happened', that it was 'only' a 'dog'.

Now, following the trace of the unseen canine carcass, one can't help but notice that 'nothingness' follows the 'dog'.[9] So, what, exactly, is the status of the 'dog' for the bourgeois subject? The 'dog' never refers to a biological reality. For the white bourgeoisie, the 'dog' functions as a pliable, empty signifier, as a wild card that is imbued with different properties at the discretion of the bourgeois individual employing it. In spite of its formal plasticity, what remains constant is that the conceptual figure of the 'dog' functions as a technology of mediation whose purpose is to promote affective desensitization.

For Vero's husband and Vero's cousin, the figure of the 'dog' is the mediating tool par excellence; the 'dog' abates all doubts. The cousin has heard the thumps on the road, not just once, not only twice, but many times; he tells himself that it must have been a 'dog', without fail; he turns a deaf ear to any critical ethical implications heard in the sound. The wealthy *salteño* invokes the 'dog' and suppresses troubling affects and sticky uncertainty – uncertain affects that might compel an examination of the unethical effects of forms of living and feeling, of relating and being. The 'dog' mediates the experience of listening for the men in Vero's life. Martel will destroy these efforts to mediate and silence the ethical issue staged in Vero's ears. Shortly after Vero's husband and Vero's cousin convince her that the body she ran over was an unseen canine one, a body will actually be found in the peripheral canal, next to the road.

Firefighters look for something that obstructs the now-flooded waterway. The type of 'matter' remains to be determined: a passer-by tells Vero's relatives that it might be the body of a calf or the body of a person. The man also comments that an animal fell in the marginal canal the previous year. This time, however, it will be determined that the body is that of a *mestizo* boy who died around the same time, or perhaps, in the exact same time in which Vero crashed into a body (or bodies) on the road.

Here one finds the deconstruction of the incident, the deconstruction of the accident: an accident cannot be thought as a singular point that can be captured in an intelligible form or as an immediacy, on the contrary, the accident in Martel is susceptible to deferral and difference; it is a spaced disturbance that resists becoming legible. But this deconstructed incident makes the question about (un-)ethical thresholds even more significant: How must one *act* in the face of uncertainty? How might affect stimulate or delay action?

Dullness: Dull affect and de-animated thought

What does one mean when one talks about 'affect'? What is the relationship between affect and sound? The fields of film and sound studies have given much attention to Gilles Deleuze's influential work on affect. Deleuze champions a tradition of affect in which affect conceptually emerges as non-representational vibration, force or resonance; affect is thought of in terms of bodily intensities that bypass cognition and signification.[10] Certain strains of sound studies have adopted this notion of affect, thus giving priority to sound as vibrational force that is registered in the body of the listener before cognitive responses crash the party.[11]

Yet, the experience of acousmatic sound joins cognition and affect. Kane suggests that acousmatic sound is *structurally* disturbing because it is constituted by a gap that prevents the unification of sonic source, cause and event at a sensory-cognitive level. Kane links anxiety to the experience of acousmatic sound.[12] For Chion, the acousmatic situation can be an 'invitation to the loss of self, to desire and fascination'.[13] By definition, the experience of acousmatic sound cannot be affectively neutral. And its affective non-neutrality derives from the (im-)possibility of knowing and not knowing.

Even though these accounts of affect are analytically opposed, for the Deleuzian often seeks to expel the operations of cognition from the embodiment of affect as sound, whereas an acousmatic-based theory of affect defends the inseparability of signification and affect, both seem to share an understanding of affect as a form of excess. Yet, Vero's affective style is anything but excessive, the thumping sounds have left her rather dull. Vero's dull affect resonates with inactivity, sleepiness and soundlessness. Does this mean that she is affectless? Of course not.

Vero's affective style can be put in the vocabulary of what Lauren Berlant terms variously 'recessive affect', 'interpassive affect' and 'flattened affect'.[14] These are all modes of underperformed emotion that show up to enact a recession from melodramatic norms.[15] According to Berlant, recessive affect resists reading and communicability, for it escapes shared codes of affective understanding. Indeed, Vero's dull affect thwarts interpretive efforts. Dull affect puts her besides herself, in a motion of de-appropriation of her 'own' self that permeates the cinematic frame with virtual non-presence. Vero becomes a hazy surface. And her dull affect frustrates readability even more than recessive affect, for, even though recessive affect expresses the non-self-evidence of the affective event, Berlant tells us that recessive affect is 'diffused yet *animated* gesture'.[16] But Vero's dull affect consist in de-animation. Hers is a new kind of recessive affect, one that not only refuses exhibition and identification: dull affect is a form of de-animated thought. The affective force of the acousmatic thumps eventuates in a loss of animation. Vero forgets that she is a dentist and that her office is her office; she forgets her own name; she forgets to wash her blonde hair. Her body forgets how to move; her mouth forgets how to speak.

Now, to say that the sonic-affective power of the thumps performs a de-animation of Vero's thought is not to say that affective de-animation and dullness amount to absence, failure or negation. That is to say, it is not the case that dull affect and de-animated thought do not do anything: these peculiar affective modes hold generative power – these affective modes do something. Once Vero embodies the sonic-affective force of the acousmatic thumps, she becomes estranged from her world. Her de-animated thought perceives what is usually known as subtlety distorted; mundane episodes appear as strange phenomena, domestic scenes and ordinary objects become infused with an uncanny symbolism. Dull affect and de-animated thought invoke unknowingness. How can one sense such a strange affective state of unknowingness?

From the visual point of view, Martel makes use of shallow focus in order to describe Vero's acts of looking once Vero has felt the force of dull affect and de-animated thought, elaborating a poetics of affective vision within the cinematic frame (see Plate 8). Vero looks at the world only to find blurriness. *Mestizo* boys appear as blurry, indistinct figures in the background; *mestizo* workers tend to be out of focus. Shapes and figures lose their edges. From the sonic point of view, Martel exploits the force of off-screen sound. The ordinary turns eerie. For instance, a creaking sound arrives to the ear. Vero's mother begs her daughter: '*don't look at them . . . the dead*'. A blurry silhouette then appears in the background of the image. There is a moment of interpretative incertitude: Is the indistinct figure a ghost? A child? Both? The domestic space of the home becomes a terrain that trembles with uncanny energy (see Plate 9). Once Vero embodies dull affect and loses the 'animated-ness' of her thought, the visible and the audible become tools for de-familiarizing the habitual.

And de-animated thought and dull affect also allow Vero to perceive the de-familiarized ordinary as a scenario of violence. Still, dull, and quiet, Vero sensually tracks and processes her world with an enhanced awareness of the brutal-but-muffled violence constitutive of the normative space of life. She experiences what Deleuze terms 'pure optical' and 'pure sound' situations, situations where one ceases being an agent [*actant*] and becomes a seer [*voyant*], situations that lengthen and distend time, situations where disconnection and suspension promote a mode of apprehension of ordinary life in all its non-ordinariness.[17] Vero's dull affect compels her to see and to listen to the brutal motions and forces that rule the sociality she belongs to. But Vero cannot act. Deleuze asserts that in pure sonic and optical situations, what is gained in an ability to see and to hear is lost in action or reaction.[18] Vero's dull affect cannot prolong itself into action. So when the body of a boy is found in the canal, sleeping-beauty Vero waits.

The metamorphosis of the incident into a non-incident, the clandestine actions taken to erase Vero's probable criminal implication in the incident, issue from the hands of the men in her life. Her lecherous brother deletes the medical records from the hospital where Vero goes after the collision to have her head examined; her incestuous cousin makes a series of phone calls to invisible figures, presumably detectives and policemen; her disloyal husband drives Vero's car out of town to get it fixed. The men take action *for* her, and their actions make the law. And the law, Mladen Dolar reminds us, 'is nothing but the movement of deferral, [the law] coincides with this perpetual movement of evasion.'[19] In Martel, the law cannot be seen, the law cannot be captured in a single frame: the law circulates from one phone call to the next, as a whisper, in an invitation to go fishing, in the small talk that is not just a performance, but that in itself is performative: the bourgeois small talk has material effects.[20]

Vero is the focal point of this choreographed universe of power, but she is an invisible, dark sun. She never delivers her *mea culpa*; she keeps silent and dyes her hair dark brown; becoming someone else is an easy task for her. The bourgeois woman obeys the obligation to keep silence: What *must* one do after one has been addressed by an ethical imperative so violent? Ammonia does it all.

There is no closure of the incident itself, no revelation of what happened. But focusing on one single point, on the 'origin' and on the facticity of such an origin is a triviality, for the mechanics of the law can be tracked down symptomatically, that is to say, by looking at the effects of effects. Effects respond to an uncertain cause: did she kill the boy? The answer: (act as if) nothing has happened. Let appearances manipulate facts – appearance *is* fact. The task consists in preserving muffled violence.

'Let the ear fall into desuetude', says bourgeois consciousness. But the ear is a hole, and a hole always lets the outside in.

Notes

1 Franz Kafka, *The Metamorphosis*, trans. and ed. Stanley Corngold (1915; New York: Modern Library New York, 1972), 30.

2 Middle of the Road, 'Soley, Soley' (1971) [audio recording], UK: RCA Victor.

3 The term was unearthed by the French composer Pierre Schaeffer in the 1950s. The 'acousmatic' is linked to a quite ancient myth according to which Pythagoras's lectures were delivered from behind a curtain, so that the listeners – the *akousmatikoi* – would attend to the meaning of the words without visual distractions. For a discussion on the similarities between the ancient myth and modern audio recording technologies, see Pierre Schaeffer, *Treatise on Musical Objects* (Oakland, CA: University of California Press, 2017), 64.

4 Michel Chion, *The Voice in Cinema*, trans. and ed. Claudia Gorbman (1982; New York: Columbia University Press, 1999), 19.

5 Brian Kane, *Sound Unseen* (New York: Oxford University Press, 2014), 224.

6 Kane, *Sound Unseen*, 148.

7 Jacques Derrida, *Positions*, trans. Alan Bass (1972; Chicago: University of Chicago Press, 1981), 4.

8 Slavoj Žižek, *Violence* (New York, Picador, 2008), 2.

9 I discuss Martel's employment of the figure of a 'dog' in her film *La ciénaga* (*The Swamp*, 2001). I focus on the manners in which the 'dog' loses its semiotic stability, provoking an anxious mode of listening in one of the diegetic characters. See Andrea Avidad, 'Deadly Barks: Acousmaticity and Post-Animality in Lucrecia Martel's *La ciénaga*', *Film-Philosophy* 24, no. 2 (2020): 222–40.

10 Gilles Deleuze puts forth a notion of 'affect' following Baruch Spinoza's ontology of immanence and Henri Bergson's writings on memory and time. Pre-personal affect forms part of Deleuze's own ontological model of pure difference or 'difference-in-itself'. For a summary of Deleuzian affect and sound, see Brian Kane, 'Sound Studies without Auditory Culture: A Critique of the Ontological Turn' in *Sound Studies* 1 (2015): 5.

11 For ontologies of affect as sonic vibration and of sound as affective, material impact, see Steve Goodman, *Sonic Warfare: Sound, Affect, and the Ecology of Fear* (Cambridge, MA and London: The MIT Press, 2012), Christoph Cox, *Sonic Flux: Sound, Art, and Metaphysics* (Chicago: The University of Chicago Press, 2018), Mack Hagood, *Hush: Media and Sonic Self-Control* (Durham, NC: Duke University Press, 2019).

12 Kane bases his model of acousmatic sound on a reading of Franz Kafka 'The Burrow' (1931). Kane suggests that the anxiety experienced by Kafka's protagonist, a mole who lives in a dark burrow, emerges because of the impossibility for determining the sonic source of a whistling sound. See Kane, *Sound Unseen*, 138.

13 Chion, *The Voice in Cinema*, 6.

14 Lauren Berlant, 'Structures of Unfeeling: *Mysterious Skin*', *International Journal of Politics, Culture, and Society* 28 (2015): 191–213.

15 Berlant, 'Structures of Unfeeling: *Mysterious Skin*', 194.

16 Berlant, 'Structures of Unfeeling: *Mysterious Skin*', 193.

17 Gilles Deleuze, *Cinema 2: The Time-Image*, trans. Hugh Tomlinson and Robert Galeta (1985; Minneapolis: University of Minnesota Press, 1989), 272.

18 Deleuze, *Cinema 2*, 272.

19 Mladen Dolar, *A Voice and Nothing More* (Cambridge, MA: The MIT Press, 2006), 165.

20 Here I refer to the 'performative' model of speech acts according to which language *does* things; language produces significant effects. For an examination of the performative model of language, see Rebecca Comay and Frank Ruda, *The Dash* (Cambridge, MA: The MIT Press, 2018), 16., J. L. Austin, *How to Do Things with Words* (Cambridge, MA: Harvard University Press, 1962), 4.

Entry 12

Buzz

Sharon Jane Mee

The meandering but insistent buzz of a fly – of several flies, a swarm even – is a sound that fuses with the white noise of a fluorescent light. The fly fills one with the horror of moving decomposition brimming with maggot offspring. Ticklish on flesh, the buzzing fly delivers a sense of disgust – something that has touched dead flesh touches living flesh. Paul Solet's 2009 film *Grace* opens with flies: flies that buzz around a fluorescent light in an RV, and over and on the naked flesh of a woman (see Plate 10). In Solet's *Grace*, the buzzing fly is the spectre not only of death but of life as a figure of sensation. This is a sensation shared with the spectator. In Darren Aronofsky's 2017 film *mother!*, a fly dying at the windowsill activates the character's (and spectator's) sensing of decomposition. The fly – in buzzing death-throes that modulate the sound to a deeper tone – ends up on its back; the bloodstain on the floorboards – left where the uninvited guests' son (Brian Gleeson) has been murdered in the house – seeps and spreads. When the messy, smelly, micro-movements of the decomposing body cannot be visualized in cinema, the sound of the fly acts as a sound affective spectre of decomposition in the horror film. Like the life that 'refuses to go away' when someone dies – the maggot offspring of the fly is the life that invades the decomposing body – this entry examines sound affects as something which 'refuses to go away' in cinema.[1] The buzzing of the fly in films such as Solet's *Grace* and Aronofsky's *mother!* acts as a spectre of decomposition, expanding the sensory field of the decomposing body in the image with sonic, moving and energetic qualities.

This entry argues for a cinema of spectres that, like the buzzing fly, is sonically and affectively undialectical. This is a cinema that has a fundamentally different relationship to death in that its sound affects are not recuperated in the image, and ultimately in the icon, that is, in the monument of the gravestone and even the monument of the filmstrip. This death is the decomposition that is always in movement, imperceptible and always the life in death in our encounter with the cinematic image. In her essay 'Critique of Silence', Eugenie Brinkema links silence 'to a dialectic of being and nothingness, plenitude and finitude, the fullness of meaning or the ground of nonmeaning from which meaning is drawn', and links what she calls the 'near inaudible' to 'pressure, tension, intensity, and force'.[2] The decomposing corpse – with flies that buzz around it – is a death not dialectical, because it is movement and 'near inaudible' sonic intensity that does not lie silent like the crypt.[3] Like the decomposing corpse, the sound of the fly – sound affective spectre of the temporal attenuation of the material state in decomposition – resides on the very edge of sonic sensibility, of the 'near inaudible' of beating wings.[4]

An encounter with the fly is an encounter with sound affects. The movement and sound of the fly is the life in death of the imperceptible movements and sounds of decomposition. The activity of the fly is specific to the activity of decomposition and is thus specific to a theorization of life that might be said to seize – to grasp – matter in what is a sound affective encounter. And this is the way that the imperceptible image of death (and indeed, decomposition) relates to the overall theme of this entry. For although death seems in every way contra a theorization of the vital signs that designate the conditions of life, in aesthetic terms, life seizes – grasps – the corpse in an encounter that is resplendent sonic and affective intensity.

I examine the work of theorists Eugenie Brinkema, Georges Didi-Huberman and Roland Barthes. For Brinkema, grief is undialectical, and in a commensurate way, my entry considers the undialecticality of death via the sound affects of decomposition. This entry speaks of grieving mothers or those who grieve for mothers – the grief of the mothers in *Grace* and *mother!* for their dead children, the grief of the mother in Koen Wessing's photograph, *Nicaragua* (1978), for her son shot down in the street, and the grief of a son, Roland Barthes, for his dead mother that he writes about in *Camera Lucida*. However, where Brinkema characterizes grief as 'a non-labor that does not profit' and, thus, as fundamentally undialectical, non-relational, and even still[5] – an *intense immobility* as Barthes writes of the photographic *punctum* that 'wounds'[6] – the decomposing and putrid corpse, while undialectical, is relational (even if abjectly so[7]), moving and transformative. My purpose is to find the fly in the micro-movements and 'near inaudible' sonic intensities of decomposition for a death whose meaning is not yet recuperated by the silence of the crypt, and in the viewer's sonic and affective encounter with the image whose meaning does not yet have the distance of preservation.[8] While my intention is to consider sound affects in film, such an undialectical sonic and affective encounter is evocative of what Brinkema writes about the photograph: 'The photograph is marked not by preservation or distance but by the immediacy of an event that has just taken place and is not yet buried.'[9] Sensation haunts the event of death: in the viewer's encounter with the image the photographic past is 'not yet buried'.[10] Such sensorial haunting has, as its constituent, the sonic and affective conditions of life. As Barthes writes: 'if the photograph then becomes horrible, it is because it certifies, so to speak, that the corpse is alive, as *corpse*: it is the living image of a dead thing.'[11] Beyond the image and our encounter with it, it is the fly that infuses the corpse with life. If the photograph, for Barthes, is 'literally an emanation of the referent' as 'radiations which ultimately touch me',[12] the fly that buzzes over the corpse is a kind of 'emanation' of the sonic and affective intensities of decomposition. The fly in film is of sound affects duplified: while the sound of the fly is an affective spectre of decomposition, film as impalpable light and sound is itself spectral sound affect. Both speak to sound affects via the energetic force of the living encounter.

Still-torture/Torture stilled

The fly is a sound affective spectre of material decomposition, but it also worries life. Marcel Duchamp's artwork of 1959 titled *Torture-morte* demonstrates the capacity of the fly to torture (see Plate 11). The artwork itself is a painted plaster foot framed within a glass-fronted wooden box. The foot's surface is bared to us like the foot of a corpse at a morgue, and upon the foot numerous (thirteen to count) flies are settled. Not only are we possessed of the idea that this must be incredibly ticklish for the owner of the foot – a sensorial torture – but further, that if it is the foot of a corpse, the sensation haunts the spectator rather than the corpse. *Torture-morte* can be translated to mean 'still-torture', like the traditional fruit and flowers of the still life (*nature morte*). *Torture-morte* is 'still-torture' as a constant or persistent torture beyond the event of death, just as it is also torture stilled, framed, as it is, within a glass case. Film will not be stilled in the same way as the sculpture. Film's flickering light, movement and sound sets in motion new affective encounters for the spectator.

Film as sensorial hauntology: The fly as sound affective spectre

In 2008, Steven Shaviro responded to papers presented at the SCMS conference in consideration of film's spectres, referring to Jacques Derrida's use of the word 'hauntology' in his book *Specters of Marx: The State of the Debt, the Work of Mourning and the New International*. Derrida's 'marvellous invention' in *Specters of Marx*, Shaviro proclaims, is the combination of *haunt* and *ontology* to 'indicate the being of that which is not manifest, not there, not present'.[13] We can note a haunting of ontological existence when Derrida writes that 'justice carries life beyond present life or its actual being-there', such that the '*living-on* (sur-vie)' to which Derrida refers is of spirits or spectres – a 'trace of which life and death would themselves be but traces and traces of traces, a survival whose possibility in advance comes to disjoin or dis-adjust the identity to itself of the living present as well as of any effectivity'.[14] For Shaviro, the haunting of ontology does not simply lie in absence or a death that teaches – disjoins or dis-adjusts – life, for, as he notes, the 'formulation is reversible; it just as well designates a continuing subsistence, or insistence, at the very heart of death and absence'.[15] In the way that Shaviro writes that 'film is itself the hauntological art *par excellence*', I argue that the 'hauntological dimension of the movies' is a 'continuing subsistence, or insistence' of sound and affect in death and decomposition. Death is encountered in the pastness of the filmstrip in which bodies once alive are now dead, although the spectator is very much alive; and also, in the films that I examine, in the life in death of the decomposition of bodies.[16] As Shaviro writes of Derrida's spectre as that which 'refuses to go away', the fly also 'insinuates itself within the very present that excludes it, or haunts the corporeality to which it cannot be reduced'.[17] That is, just as the fly insinuates susurration and wriggling disgust for the corpse – the object of which must be radically excluded[18] – into death, the fly 'refuses to go away'.[19] Like the uninvited guests in Aronofsky's *mother!* that 'refuse[. . .] to go away', the fly is always an uninvited guest to the corpse.[20]

In Solet's *Grace*, the fly does point not simply to a dead body but to life as a figure of sensation. The film begins (and ends) with a question: Is the baby dead or alive? A baby is dead inside its mother's womb at eight months. The mother, Madeline Matheson (Jordan Ladd), carries the child to term and gives birth naturally, knowing that the child will be stillborn. Holding the newborn child in her arms, it emits a cry. It is alive, although it has a smell that attracts flies and will only sup on blood. While the baby, Grace (Tenai Cam Measmer), is sleeping, a fly buzzes around her face – into her nostril and out again. Madeline proceeds to hang fly paper from the ceiling to catch the flies. As a mass of flies gather on the mosquito netting over the crib, Madeline pins more fly paper to the ceiling until the room is an obstacle course of hanging fly paper. (Does film hope to capture and 'embalm'[21] these flies within its emulsion-coated filmstrip in the same way that the fly paper aims to catch flies on its sticky surface?). Like the insistent buzzing of the fly that 'refuses to go away', sensation also 'refuses to go away' in cinema.[22]

The sound of the fly operates as an affective spectre of the materiality of the corpse, like the sound and movement of the image that operates as an affective spectre of the materiality of the cinematic apparatus. It is squirming life that haunts the baby body – flies and all – in *Grace*, just as it is life that haunts the image. Beyond image, Davina Quinlivan argues that the kinaesthesia of sound lends itself to 'material hauntings'. She writes: 'Sound, in particular, generates a current of physical energy felt by the viewer while remaining ambiguous, fleeting, only partially grasped as lost moments in time.'[23] The sibilation of the fly, ambiguous and fleeting, at the same time constant, arouses feelings of repulsion, disgust and irritation. For films made in the post-cinematic age such as *Grace*, the buzzing fly has a double effect. The sound and affective tactility of flies on flesh fuse with the buzzing aurality and flickering pixels of the image. The digital image, composed of transitory pixels and sounds, has the force of an energetic field. In its moving pixel terrain, the digital image has the behaviour and expression of a 'fleeting' but 'physical energy'.[24] From both the materiality of bodies on screen and the cinematic apparatus emanates the physical and energetical forces of sound and light, such that this 'film' operates as an energetic field.

The insistence and persistence of the stain – and of the fly – in Darren Aronofsky's mother! *(2017)*

A fight ensues in Aronofsky's film *mother!* when the sons of the uninvited guests (Ed Harris and Michelle Pfeiffer) arrive at the house. The older brother (Domhnall Gleeson) demands to know whether his dying father knew what he signed when he signed his will. The father explains that it is a trust that they must make a decision as a group before any money can be spent. The younger brother exclaims that their father is dying, and all his brother wants is money. The brothers fight. The younger brother spits in the older brother's face, who throws a chair which splinters against the wall. The older brother uses a doorknob to hit his younger brother until the younger brother lies still with blood pooling around his head. In a flurry of activity – shot with a handheld camera that follows the erratic movements of mother (Jennifer Lawrence) around the house as we might follow the erratic movements of a fly – Him (Javier Bardem) calls for towels, asks mother to call for the father of the son – a doctor – who has retired to another room, and then Him, with the help of the father, carries the younger brother out the door, while mother pleads 'don't leave me here'.[25] All is quiet in the house. The buzz of a fly dying at the windowsill disturbs the moment and, by its frenzied death, activates a sensing of decomposition, foreshadowing the dissolution of events that will lead to mother's death (see Plate 12). Mother prepares to mop the blood from the floor.

If the photographic image acts as a monument to death, Didi-Huberman's essay 'The Index of the Absent Wound (Monograph on a Stain)' is instructive in that it considers a stain – the stain on the shroud of Turin. Didi-Huberman describes a viewer who gazes at the surface of the photographic negative of the shroud that the viewer is allowed access to in Turin, and 'desires to see' beyond the material; to see what is absent – the wounds of Christ – in the stain.[26] The stain, however, resists sublation, as Didi-Huberman writes:

> It doesn't seem to lend itself to being raised up (in the sense of the dialectical *Aufhebung*) into something figurative; it seems to defy comprehension as a recognizable image. It says nothing about the economy of its support (which would at least establish the hypothesis of a luminous-negative index). It seems to exist only in terms of its tonal variations, only as an effect of its support. Yet the tonal variations of the fabric have no precise limits, sequence, or articulation. It seems to exist, therefore, only as the uncertain effect of something as undifferentiated background. Between the *spatium* (the background in question) and the pure surface, this stain reveals itself only in the precarious opening of the becoming visible.[27]

The bloodstain on the floor in the film *mother!* also 'reveals ... the precarious opening of the becoming visible'.[28] This 'becoming visible' is activated by a 'desire to see' that scrutinizes the material to discover *something*.[29] For Didi-Huberman, the 'desire to see' is a desire to see a body – the body of Christ – in the stain as an index of His wounds. Michel Chion also describes a '*desire to see what is going on*' in Alfred Hitchcock's *Psycho* (1960).[30] For Chion, it is the offscreen voice which, when all is revealed, creates the 'desire to see' 'the incestuous marriage between the mother's voice and Norman's body'.[31] Acousmatic cinema is characterized by sound detached from a body, and indeed, for *Psycho*, what Chion calls the 'impossible couple of body and voice'.[32] The acousmêtre in the form of the voice of the mother in *Psycho* is typified by an 'impossible embodiment' because of the impossibility 'to return the dead to life'.[33] Chion writes: 'Ultimately the voice has not found a body to own it and assign it a place'.[34] D. Ferrett argues that, in fact, Chion makes 'powerless' the 'sonic powers of the mother'; that is, 'in claiming to raise the sonic powers of the mother, upon closer analysis these powers are subsumed under a masculinist scopic regime, one that attributes powerlessness to lacking female bodies, once the lack is "discovered"'.[35] In opposition to the offscreen voice of the primary creator of the child in the womb, mother in *mother!* is visually (and vocally) present – she persists – and is followed by the camera throughout her enfleshed flurry of activity where body and (camera) movement become entangled against the backdrop of the affective material world. She is nevertheless vocally impotent and unable to sway her husband in his opinion of the uninvited guests. In this sense, *mother!* retains Chion's masculine (scopic) regime under which female subjectivity is found lacking through mother's aural register of 'discursive impotence'.[36] However, where, through Ferrett's analysis of Kaja's Silverman's critique of the female 'striptease' that takes place for Chion's acousmêtre, Ferrett argues that for Chion the 'mystery of the mother's body [must] be maintained, because in order to believe in these sonic powers her lack must be kept out of sight',[37] mother in *mother!*, followed by the camera, opens spaces and sensations as if to – and from – her body (her mouth?).

Writing about the spectral excess of bodies, Shaviro describes the vocalizations in Robert Aldrich's *Kiss Me Deadly* (1955): 'the excess of these sounds implies both a violent literalization, a hyper-materialization of affect, and, at the same time, an irreducible excess of spectrality or spirit, as if the interiority buried in the depths of bodies were also that which renders them impalpable, forever beyond our grasp'.[38] While the sounds of breathing, crying and muttering in *Kiss Me Deadly* are sounds that issue from the depths of bodies, the buzzing of the fly – a 'violent literalization' of death and 'hyper-materialization of affect' – is, by its sonic intensities and droning durations, a spectre of the micro-movements of decomposition 'buried in the depths of bodies'.[39] Far from an 'impossible couple',[40] the fly that dies at the windowsill in *mother!* is an excessive spectre of sound and affect for the materiality of bloodstain, floorboards and wall. The buzzing fly in *mother!* is not the spectre of an absent body; a fly always finds a body (in the way that the fly that lands on Norman's hand in the final scene of *Psycho* has found a body). Rather, the fly speaks to an excess of life – and of body – in death/decomposition. Thus, in violent opposition to 'the acousmêtre [that] "haunts" at the "edge of the filmic frame"',[41] I want to argue that the spectrality of the sound of the fly announces a body not yet buried, although unlike the 'exhumed and stuffed'[42] corpse of Norman's mother for which vocal sounds cannot find a body, the sound of the fly in *mother!* is an excessive spectre of the imperceptible noises of the decomposition of the house and life within its environs.

The beating of a fly's wings that produces the buzzing sound, erratic and near, now far, resides on the edge of the imperceptible (decomposition resides in this sonically imperceptible domain). This suspension at the limit of audibility is what Brinkema calls 'near inaudibility'.[43] For Brinkema, the 'regime of near inaudibility . . . advances pressure and intensity and force in place of presence and absence'.[44] For the mothers in *Grace* and *mother!*

this is the 'near inaudibility' of the tiny, beating wings of flies, as much as the 'pressure, tension, intensity and force' of grief and disgust that attends death and decomposition.[45] Indeed, the 'pressure, tension, intensity, and force' of grief is also that of Barthes's photographic *punctum* that 'wounds' its spectator.[46] Just as Didi-Huberman's stain defies 'comprehension as a recognizable image'[47] bearing witness to an absent wound for stains 'neither imperceptible nor yet perceptible as figures',[48] the sound of the fly bears witness to the vital operations of decomposition in which the silent monument of death does not figure. Unlike the 'desire to see' wherein one hopes to find the presence of meaning in figurative form, the desire to hear exposes the emergence of intensities and durations of tension in the leaning forward towards materiality and its micro-movements. Brinkema writes about the treatment of aurality in Brad Anderson's 'Sounds Like' (televised 17 November 2006): 'The aural order is a dynamic one addressed to material vitality, demanding the absence of sound at the simultaneous death of movement. The death in the film fails this demand on both counts, the body noisy in decay and frenetic in rot's progress.'[49] Similarly, the sound of the fly is an address (addressing the viewer) to the movement in decomposition. The droning buzz of the fly draws out durations of intensity, which, like decomposition, is not still for the silence that we demand of death.

For Brinkema, the distinction between silence and 'near inaudibility' is a distinction between stillness and duration: 'The violence of silence is to being, in order to render that kinetic stillness; the violence of near inaudibility is to form, sustaining the pressure of a duration – creating that "space of time" for the sensation of extreme quiet to manifest.'[50] Within Brinkema's distinction is the joke affirmed by Duchamp's *Torture-morte*: sensation won't be stilled. The sensations that torture the still dead – a corpse that is dead-still – are the spectator's in their apprehension of the flies on Duchamp's artwork. Just as Brinkema relays how, in the original story by Edgar Allan Poe that was adapted for the television episode 'Sounds Like', Larry finds no reprieve after murdering his wife – 'The irony of it is that she's louder dead than she ever was alive. . . . He hears it in the awful drone of the flies and the soft hiss of her putrefying organs, but more than that, it's revealed in the sounds of his own treacherous body'[51] – the sensations of *Torture-morte* are those of the spectator's own 'treacherous body'.[52] There is still life in death.

Sound affects are activated in each new encounter with the image. This is the 'refus[ing] to go away' of sensation.[53] As Shaviro writes: 'Something that has died, something that is in the past, nonetheless refuses to go away.'[54] The bloodstain also 'refuses to go away' in *mother!*[55] Mother mops up the blood and scrubs at the stain on the floorboards, even as the blood seems to have seeped into the grain of the wood and between the floorboards, leaving a gash of crumbling wood. Even when the stained floorboards are replaced, the stain reappears on the floor mat, and later, reappears on the floorboards. Sound is even more persistent than this stain. In *Grace*, the buzzing of the fly is so persistent a sound, existing at the limits of audibility, such that even once the fly has moved away the sound of buzzing and the feeling of disgust persists for the spectator. Thus, the intense durations of languid buzzing, and erratic, high-pitched whizzing to the deep-toned death-throes of the fly, in the way that Brinkema describes the force of duration of the 'near inaudible', 'sustains its simultaneous rising out of and disappearance into another scene, without orientating itself in relation to that scene . . . What is left, as at the scene of some indeterminate crime, are the traces of the affective possibility of what sound might become.'[56] If the bloodstain in *mother!* is an index of violence having taken place – a marker for the scene of a crime – the buzzing fly is an affective possibility in the *becoming* 'of some indeterminate crime'.[57] The 'indeterminate crime' may well be the 'near inaudible' sounds of decomposition that disrupt the silence of the grave.[58]

An 'ontological fold': Dynamism and an openness to relations

Each cinematic event is an affective encounter with image and sound for the viewer. An affective encounter – to be 'here' and 'now'[59] – with the photograph may be understood by what Barthes describes in the *punctum*: 'this wound, this prick, this mark made by a pointed instrument'.[60] For Barthes, the *punctum* 'also refers to the notion of punctuation'.[61] He writes, 'the photographs I am speaking of are in effect punctuated, sometimes even speckled with these sensitive points; precisely, these marks, these wounds are so many *points* . . . *punctum* is also: sting, speck, cut, little hole – and also a cast of the dice'.[62]

Flies, by their affect, are insistent speck that punctuate surface. And indeed, an example given by Barthes of the *punctum* is Koen Wessing's photograph of a weeping mother carrying a sheet towards a child's corpse (see Figure 12.1).[63] What strikes Barthes about this photograph is 'the corpse's one bare foot, the sheet carried by the weeping mother (why this sheet?), a woman in the background, probably a friend, holding a handkerchief to her nose'.[64] What strikes me about this photograph is that the white sheet covering the corpse appears to have flies – they seem too regular in size to be drops of blood or marks on the surface of the photograph – settled upon its white surface. Imagine little holes made by a pointed instrument, punctuating the corpse's

Figure 12.1 Flies on a sheet-covered corpse in Koen Wessing's *Nicaragua* (1978) © Koen Wessing/Nederlands Fotomuseum.

existence, and wounding, pricking, marking, stinging, specking, cutting and casting the die for the spectator.[65] Catalysed through Didi-Huberman's 'desire to see' – which, like the shroud of Turin, to 'see' wounds beyond the surface of the photograph in those specks on the white sheet – the flies 'take[. . .] on the form of a dizzying spiral' which 'involves an entire constellation of . . . phantasms'.[66] The wounding of the viewer by *punctum* is both exteriority (the fall into surface; here, the surface of the sheet covering the corpse of the boy) and interiority (the fall into sensation; here, a 'dizzying spiral' that goes beyond the surface of the sheet through flies that mark the decomposition of the boy's body to mark also the 'dizzying spiral' of the mother's grief).

While the fall into surface may be a fall into the photographic material surface as 'a carnal medium, a skin I share with anyone who has been photographed', the fall into sensation is a fall into impalpable light.[67] This is the spectre of rays of light in the Winter Garden Photograph of Barthes's mother that Brinkema considers, as well as of the photographic negative of the shroud of Turin illuminated from behind. Writing about the rays of

light that 'touch' the bodies in the photograph that also 'touch' the viewing
body, Brinkema contends: 'What photography thus *graphs* is not just an
image enabled by light but that image captured as light in its lightness.
The significance of figuring photography as a medium without mediation,
wherein rays from the photographed body directly touch a viewing body, is
a profusion of corporeal figurations, as if touch signaled the final death of
dialectics.'[68] Thus, Brinkema suggests that the 'return of the dead that Barthes
sees in all photographs is not a figural return but an ontological fold'.[69] In a
similar ontological folding, the *punctum* has a dynamical connection – at the
point of contact – by which the viewer is manifolded with the photograph and
the world, indeed, the world of death/decomposition and grief.[70] In the films
with which I am concerned, it is not simply the corporeal figurations of flies
on flesh that may 'touch a viewing body',[71] such corporeal figurations may
be extended to sound affects in vibrating reverberation of air and affective
intensities of sound. The sound of the fly, like the photograph, is a complex
sensory-affective experience that enfolds the spectator.

   Touch is deeper than surface in *mother!* As of the blood that spreads and
seeps into floorboards, mother senses the life of the house when touching
the walls inside the house, visualized in a beating heart which shrinks and
blackens. Life is felt in the quivering vibrations of sensation felt in the walls
of the house – and later in the womb of mother – and settled by a bright
yellow effervescent powder dissolved in water for drinking. For Quinlivan,
the 'beating wings of insects and branches softly falling to the ground' in
Lucile Hadžihalilović's film *Innocence* (2004) 'emphasizes the difficulty of
placing oneself at a remove from nature, re-inscribing a bodily connection
to the space and to matter that is not entirely perceptible'.[72] Equally, the
buzzing of the fly in *mother!* emphasizes the 'not entirely' perceptibility
of decomposition that has sonic, moving and energetic qualities. Film, by
're-inscribing a bodily connection to the space and to matter that is not
entirely perceptible',[73] has the potential in *mother!* of opening the human
to the non-human of the house, and indeed, to the non-human of the fly.
Such a 'becoming-intense, becoming-animal'[74] has the potential of affect that
Brian Massumi writes about as the 'passing of a threshold, seen from the
point of view of the [body's] change in capacity'.[75] Where bodies pass beyond
thresholds is in their opening to sonic and affective worlds.

   Openness is closer to Barthes's *punctum* (as of an open 'wound') than it
is to the indifference found in the 'unconcerned desire' of *studium*,[76] making
the sensory-affective experience anything but banal. Indeed, the buzzing fly
is anything but banal, for it resides on the very edge of the cataclysmic: of the
sound affective imperceptibility of life in death. The *punctum* is an impact
upon the senses by which the subject is opened to the world on the one hand,

and a sensory manifold in the dynamical relation between the subject and the world on the other. Thus, just as Barthes writes that 'to give examples of *punctum* is, in a certain fashion, to *give myself up*,'[77] so too the encounter with the fly is to *give myself up* in the sense that 'I' am open to the world. In the cinematic encounter 'I' *give myself up* to affect (grief, disgust, irritation). In *mother!*, it is the mother who *gives herself up*, and indeed, gives everything to love. In the final to last scene of the film, Him implores: 'I need one last thing.' Mother responds: 'I have nothing left to give.' Him answers: 'Your love. It's still there, isn't it?' Mother responds: 'Go ahead, take it.' Him reaches into mother's burnt and charred chest cavity and, as she exhales and expires, her body crumbling into ash, within his hands he holds the veined crystal of her heart that will bring life and everything back to the beginning.

The kind of cinematic encounter with death that I describe is not of a dialectical silence that demands an absence of being (of the corpse) or emanations from bodies that demand figurative presence (the kind of certainty that we demand of the stains on the shroud of Turin). Rather, as an event, cinema communicates materiality via the excessive spectrality of sound affects. The buzzing fly opens out enfleshed life through ontologically enfolded sonic and affective relations, just as film does. The effect of the buzzing fly is not just of the material body that it touches, but the dynamical connections that it makes. The horror of the buzzing fly, in this sense, is that it allows the spectator to enter into new relations with the corpse – the undialectical image is energetically and materially relational. For film to abandon its monuments – dialectical presence and absence that recuperates meaning both in the image and in the filmstrip – in exchange for sound affective spectres of materiality is another way of disrupting the autonomy of the sign and relaunching it as a thing for the play of affect and the sonic frenzy of decomposition. New relations are formed in cinema via sound affective spectres: the sound of the fly opens out sensorial relations. An undialectical and corporeal sign – a corpse with an openness to relations – puts the viewer in a putrid state of affairs.

Notes

1 Steven Shaviro, 'Untimely Bodies: Towards a Comparative Film Theory of Human Figures, Temporalities and Visibilities', *SCMS* Conference, 2008, http://ftp.shaviro.com/Othertexts/SCMS08Response.pdf (accessed 27 November 2020).

2 Eugenie Brinkema, with composition by Evan Johnson, 'Critique of Silence', *Differences: A Journal of Feminist Cultural Studies* 22, nos 2–3 (2011): 213.

3 Brinkema, 'Critique of Silence', 211. In Brinkema's essay even this silence of the crypt is not possible, for she relays how in Edgar Allan Poe's story 'Silence – A Fable': 'The earth at Larry's son's grave vibrates with noise; pulling up the wriggling life underneath it, confirming noise even in and after death, Larry's impossible silence of plenitude is devastated. The agitation of audibility surrounds him totally and inescapably' (Brinkema, 'Critique of Silence', 223).

4 Brinkema, 'Critique of Silence', 223.

5 Eugenie Brinkema, *The Forms of the Affects* (Durham, NC and London: Duke University Press, 2014), 76.

6 Roland Barthes, *Camera Lucida: Reflections on Photography*, trans. Richard Howard (1980; London: Vintage, 2000), 49.

7 In *Powers of Horror: An Essay on Abjection*, Julia Kristeva writes about the abject of the corpse: 'The corpse (or cadaver: *cadere*, to fall), that which has irremediably come a cropper, is cesspool, and death; it upsets even more violently the one who confronts it as fragile and fallacious chance. A wound with blood and pus, or the sickly, acrid smell of sweat, of decay, does not *signify* death. In the presence of signified death – a flat encephalograph, for instance I would understand, react, or accept. No, as in true theatre, without makeup or masks, refuse and corpses *show me* what I permanently thrust aside in order to live' (Julia Kristeva, *Powers of Horror: An Essay on Abjection*, trans. Leon S. Roudiez (1980; New York: Columbia University Press, 1982), 3).

8 Brinkema, 'Critique of Silence', 211.

9 Brinkema, *The Forms of the Affects*, 78–9.

10 Brinkema, *The Forms of the Affects*, 79.

11 Barthes, *Camera Lucida*, 78–9.

12 Barthes, *Camera Lucida*, 80.

13 Shaviro, 'Untimely Bodies'.

14 Jacques Derrida, *Specters of Marx: The State of the Debt, the Work of Mourning and the New International*, trans. Peggy Kamuf (1993; New York and London: Routledge, 2006), xx.

15 Shaviro writes that this neologism 'works better in French, where *hantologie* and *ontologie* are pronounced nearly the same' (Shaviro, 'Untimely Bodies').

16 Shaviro, 'Untimely Bodies'.

17 Shaviro, 'Untimely Bodies'.

18 Kristeva, *Powers of Horror*, 2.

19 Shaviro, 'Untimely Bodies'.

20 Shaviro, 'Untimely Bodies'.

21 André Bazin, *What Is Cinema? Volume 1*, trans. Hugh Gray (1967; Berkeley, Los Angeles and London: University of California Press, 2004), 14. The desire to *immobilize* death in practices that have their origin in magic or religion, Bazin writes, demonstrates a primitive relation to death that depended upon the 'continued existence of the corporeal body' as was seen

in ancient Egypt's very literal mummification of the body (Bazin, *What Is Cinema?*, 9).

22 Shaviro, 'Untimely Bodies'.

23 Davina Quinlivan, 'Material Hauntings: The Kinaesthesia of Sound in *Innocence* (Hadžihalilović, 2004)', *Studies in French Cinema* 9, no. 3 (2009): 218.

24 Quinlivan, 'Material Hauntings', 218.

25 I use the character names mother and Him as listed in the film's credits.

26 Georges Didi-Huberman, 'The Index of the Absent Wound (Monograph on a Stain)', *October* 29 (Summer 1984): 63.

27 Didi-Huberman, 'The Index of the Absent Wound', 65.

28 Didi-Huberman, 'The Index of the Absent Wound', 65.

29 Didi-Huberman, 'The Index of the Absent Wound', 63.

30 Michel Chion, *The Voice in Cinema*, trans. and ed. Claudia Gorbman (1982; New York: Columbia University Press, 1999), 141.

31 Chion, *The Voice in Cinema*, 149.

32 Chion, *The Voice in Cinema*, 149.

33 Chion, *The Voice in Cinema*, 140, 150.

34 Chion, *The Voice in Cinema*, 149.

35 D. Ferrett, 'The Black Hole Song of Unsounding Mothers', in *Dark Sound: Feminine Voices in Sonic Shadow* (New York: Bloomsbury Academic, 2020), 116.

36 Kaja Silverman, *The Acoustic Mirror: The Female Voice in Psychoanalysis and Cinema* (Bloomington and Indianapolis: Indiana University Press, 1988), 67.

37 Ferrett, 'The Black Hole Song of Unsounding Mothers', 117.

38 Shaviro, 'Untimely Bodies'.

39 Shaviro, 'Untimely Bodies'.

40 Chion, *The Voice in Cinema*, 149.

41 Ferrett, 'The Black Hole Song of Unsounding Mothers', 116.

42 Chion, *The Voice in Cinema*, 149.

43 Brinkema, 'Critique of Silence', 224–31.

44 Brinkema, 'Critique of Silence', 223.

45 Brinkema, 'Critique of Silence', 213.

46 Brinkema, 'Critique of Silence', 213; Barthes, *Camera Lucida*, 26–7.

47 Didi-Huberman, 'The Index of the Absent Wound', 65.

48 Didi-Huberman, 'The Index of the Absent Wound', 66.

49 Brinkema, 'Critique of Silence', 224.

50 Brinkema, 'Critique of Silence', 224.

51 Brinkema, 'Critique of Silence', 223; quoting from Edgar Allan Poe and Hervey Allen, 'Silence – A Fable', in *The Complete Tales and Poems of Edgar Allan Poe* (1837; New York: The Modern Library, 1938), 125.

52 Brinkema, 'Critique of Silence', 223; quoting from Poe and Allen, 'Silence – A Fable', 125.

53 Shaviro, 'Untimely Bodies'.

54 Shaviro, 'Untimely Bodies'.
55 Shaviro, 'Untimely Bodies'.
56 Brinkema, 'Critique of Silence', 213, 229.
57 Brinkema, 'Critique of Silence', 229.
58 Brinkema, 'Critique of Silence', 229, 213.
59 Barthes, *Camera Lucida*, 5.
60 Barthes, *Camera Lucida*, 26.
61 Barthes, *Camera Lucida*, 26.
62 Barthes, *Camera Lucida*, 26–7.
63 Barthes, *Camera Lucida*, 23.
64 Barthes, *Camera Lucida*, 23–5.
65 Barthes, *Camera Lucida*, 27.
66 Didi-Huberman, 'The Index of the Absent Wound', 63.
67 Barthes, *Camera Lucida*, 81.
68 Brinkema, *The Forms of the Affects*, 83.
69 Brinkema, *The Forms of the Affects*, 83.
70 No doubt, this is not the silence of mourning for his dead mother that
    Barthes hopes for when looking at the Winter Garden Photograph. As
    Brinkema writes: 'Mourning's language (loss, absence, nothingness, death)
    is fitting for the figuration of silence as terminus, even if an impossible one'
    (Brinkema, 'Critique of Silence', 218).
71 Brinkema, *The Forms of the Affects*, 83.
72 Quinlivan, 'Material Hauntings', 220.
73 Quinlivan, 'Material Hauntings', 220.
74 In *A Thousand Plateaus: Capitalism and Schizophrenia*, Gilles Deleuze and
    Félix Guattari suggest that 'becoming-intense, becoming-animal' is the way
    affect 'throws the self into upheaval and makes it reel' (Gilles Deleuze and
    Félix Guattari, *A Thousand Plateaus: Capitalism and Schizophrenia*, trans.
    Brian Massumi (1980; Minneapolis and London: University of Minnesota
    Press, 1987), 240).
75 Brian Massumi, *Politics of Affect* (Cambridge: Polity Press, 2015), 4.
76 Barthes, *Camera Lucida*, 27.
77 Barthes, *Camera Lucida*, 43.

Part IV

Beyond audibility

Inaudible tremors

Luke Robinson

In this entry, my close analysis and discussion of three D. W. Griffith films –
the 1909 short Biograph film *The Sealed Room*, the longer 1912 Biograph film
An Unseen Enemy and the 1919 United Artists' feature film *Broken Blossoms*[1]
– give rise to a cinema concept, a sound affect, which I call inaudible tremors.
Griffith's *The Sealed Room* is an adaptation of Edgar Allan Poe's 1846 'The
Cask of Amontillado'; however, this short story by Poe is also influential for
the other two films I am discussing. In 'The Cask of Amontillado', a drunken
man is bricked up in an alcove by the story's narrator. Just before the narrator
completes his task, he takes a moment to cast a light into the space that,
because of the wall, is now dark. In response to the light being shone into the
alcove, the narrator says: 'A succession of loud and shrill screams, bursting
suddenly from the throat of the chained form, seemed to thrust me violently
back.'[2] In Poe's short story, the wall could be considered as a type of screen
– one that is sealing and concealing the narrator's victim from the outside
world. Breaking through this screen is the victim's scream, one whose force
is such that the narrator feels he is being thrusted back from the wall he
is building. *The Sealed Room*, *An Unseen Enemy* and *Broken Blossoms* also
feature scenes of women and men being confined to small spaces. In *An
Unseen Enemy* and *Broken Blossoms*, female protagonists scream and cry out
when they come face-to-face with their potential death, and, in the case of
An Unseen Enemy, there is also a male protagonist who cries out when he
thinks his two sisters might have just been shot dead. In these two Griffith
films, the cinema screen acts as a fourth wall and contributes to confining
the protagonists within a claustrophobic space. As these films are silent, the
screams and cries of the protagonists cannot be heard by the viewer who is
watching these protagonists on the cinema screen. Nonetheless, the force of
the screams and cries projected from the mouths of the films' protagonists
makes the cinema screen vibrate. It is these vibrations of the screen that I call
inaudible tremors.

The vibrations of the cinema screen caused by the projections of screams and cries from the protagonists of *An Unseen Enemy* and *Broken Blossoms* can be likened to the tremors that ripple through the earth before, during and after an earthquake. Since the tremors of these films are produced by screams and cries that cannot be heard by any viewer's physical ear, these tremors are inaudible. I argue that a study of the inaudible tremors of these three films directed by Griffith not only tells us something about Griffith as a director but also gives us insights into how an inaudible sound can be a sound affect. In this co-edited collection, there are numerous approaches to sound, affect and *sound affect*. In this entry, I argue that a sound affect is the following: it is a sound that is in the process of being actualized – a sound that is actualizing – rather than being a fully actualized sound.

Beginning with a discussion of *The Sealed Room*, the approach I take in what follows is largely descriptive. I contend that description can help us be attentive to what would otherwise go by unnoticed. Description can also help us give some 'form' to something that is without form or which is formless.[3] My approach to Griffith's films is also influenced by Gilles Deleuze who argues at the end of *Cinema 2: The Time-Image*: 'A theory of cinema is not "about" cinema, but about the concepts that cinema gives rise to and which are related to other concepts corresponding to other practices.'[4] Deleuze's own approach is not descriptive; however, inspired by his approach, I demonstrate how descriptions of films can enable cinema concepts to arise from the films themselves. For me, as a creative practitioner and film theorist, method and theory are deeply intertwined and, in this entry, description is a method for actualizing inaudible tremors as sound affect.

Actualizing inaudible tremors as sound affect

A cinema screen is generally understood as a surface for the projection of light that originates from outside the diegesis of the film. However, by looking closely at scenes and shots of screams and cries from *The Sealed Room*, *An Unseen Enemy* and *Broken Blossoms*, we can learn how forces, including sonic forces, are projected towards the cinema screen from within the diegesis of these films. In Griffith's *The Sealed Room*, The King (Arthur V. Johnson) bricks up two unsuspecting adulterers behind a closed wall, silencing their cries and screams as they struggle for breath.[5] In *An Unseen Enemy*, two orphans, The Older Sister and The Younger Sister played respectively by Lillian Gish and Dorothy Gish, are locked in a room while The Thief (Harry Carey) and the orphans' maid (Grace Henderson) attempt to crack a safe that contains the orphans' inheritance. Jacques

Aumont argues that the framing of each shot of *An Unseen Enemy* coincides with the spatial parameters of the room being filmed.[6] Crucially for the film's suspense, in *An Unseen Enemy* there is a locked door with a keyhole and a larger, inexplicable but appropriately gun-sized hole that connects the two separate rooms: one room that the orphans are in and another room where the safe is located. In *An Unseen Enemy*, tension is generated by the presence of the maid's loaded gun that is sometimes poked through the hole in the wall that separates the two rooms. When the gun initially protrudes through the hole an intertitle says: 'They silence the children while they work.' Aumont argues that the holding of the gun through the hole in *An Unseen Enemy* is a literal demonstration of the 'threat of death' that is 'always implicitly and metaphorically proffered from beyond the frame'.[7] In the earlier film *The Sealed Room* the tension of the film is visually represented by a wall being built that is imprisoning its two victims as they carry out their affair. In *The Sealed Room*, this tension has a visualized end point – the moment that the room is completely sealed. Differently in *An Unseen Enemy*, suspense is created by the maid and her loaded gun. In this latter film the gun represents the threat of death that, as the intertitle says, is silencing the two sisters. The threat of death, the gun, traverses the edge of the frame into the room that acts as a prison for the two orphans. In the film the gun in close-up initially points towards and threatens the viewer. If the screen is the fourth wall, and if the gun is a visual representation of an affect – suspense, then we can imagine how an affect, such as suspense, could also potentially traverse the boundary – the screen – that separates a film's mise en scène from a viewer in the cinema. Like the wall in Poe's short story 'The Cask of Amontillado', the screen could be breached when a sonic force is projected through it.

At one point in *An Unseen Enemy*, the gun that is threatening the two sisters withdraws from the hole when the maid decides to have a drink. The withdrawal of the gun creates a temporary safe passage between the orphan's hiding place, in the far-left-hand corner of the room, and a telephone that is positioned in the foreground. The Younger Sister takes advantage of the withdrawal of the gun and slowly creeps over to the telephone. On reaching the telephone, The Younger Sister finds that her brother (Elmer Booth) is at the end of the line. Prior to The Younger Sister picking up the telephone receiver, the film cuts between alternating shots of the two rooms of the house that are separated by the wall. Once the younger of the two sisters picks up the telephone receiver, the shots intercut between her and her brother on the other end of the telephone. We cannot hear the conversation that is taking place on the telephone, we can, however, see her brother and her mouths are moving. Michel Chion argues that films that do not synchronize voices to talking mouths on screen are a type of 'deaf cinema'.

As he says, in the films of 'deaf cinema', characters frequently move their lips, but the viewer cannot hear what the characters are saying.[8] We cannot hear what the two siblings are saying in this telephone scene of *An Unseen Enemy*, but the duration of each shot of the younger orphan and then of her brother on the telephone has a direct correspondence to the length of time it takes for them to speak. In this regard the sequence demonstrates Béla Balázs's argument that 'The actor in the silent film spoke in a way intelligible to the eyes, not the ears.'[9] The growing tension of the telephone scene is linked to the ability of the younger of the two sisters to convey the nature of her situation before the gun that reappears through the hole and cuts her off. It is the visible length of time it takes each of the protagonists to say what they need to say that creates the suspense of this sequence. In this scene, affect – suspense – takes the form of a sound – the voices of the protagonists – that the viewer cannot hear.

Suspense is often produced because of an expectation of a future danger.[10] In *An Unseen Enemy*, the gun returns once the burglar and the maid realize that the younger orphan is on the telephone. The younger orphan is then alerted to the returning gun by the cry of the older orphan. The expression of their fear in this sequence is played out largely through their mouths that appear to be gasping for breath. In *The Sealed Room* The King laughs as 'his favored one' (Marion Leonard) and The Minstrel (Henry B. Walthall) gasp for breath once they find themselves trapped in their new cell.[11] In *An Unseen Enemy*, the two sisters are visibly gasping for breath, a gasping that is simultaneous to the spiralling tension generated by the wavering arm that wields a loaded gun. After the younger orphan drops the telephone, the film cuts to her brother calling out to her. The next shot is of the room that features the orphans and we see the gun firing (see Figure 13.1). The film then returns again to the shot of the orphan's brother; he is now calling out in alarm (see Figure 13.2). In this sequence the hinge or 'fulcrum' of the shot shifts from the wall, between the shots of the burglars and the orphans, to the telephone line, between The Brother and the younger of the two sisters, an observation that is also made by Aumont.[12] However, while Aumont highlights that the use of the parallel montage shifts to one organized around the telephone line, he ignores that there is a shift of focus from the hole in the wall – which is a spatial divide – to the opening and closing mouths of the sisters and their brother. The hole in the wall that forms the edge of the frame is representative of the threat of death. But it is also, albeit an unlikely, a means of escape since the edge of the film's frame is a threshold for movement, including sonic forms of movement, to travel in multiple directions, often simultaneously. It is the orphans' desire to escape that is being represented with the unheard cries of their opening mouths.

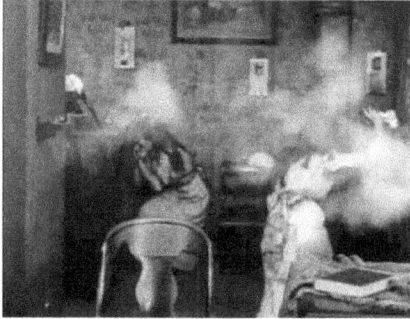

Figure 13.1 A gun is fired, and the room is filled with smoke as two orphans cover their ears in D. W. Griffith's *An Unseen Enemy* (Biograph Company, 1912).

Figure 13.2 The orphan's brother cries out into a telephone in D. W. Griffith's *An Unseen Enemy* (Biograph Company, 1912).

For the viewer, because of the parallel montage technique, the shot of the gunfire is simultaneous with the cry of The Brother at the end of the line.[13] This simultaneity of the gunfire and The Brother's cry is despite the fact that the communication line to his sister has now been cut off since, as previously mentioned, she was forced to drop the telephone receiver. In *The Voice in Cinema*, Chion cites Alfred Hitchcock's idea that if a face is visible in a film's shot, then the viewer will look at the face first. Chion argues the voice is the same, that when there is a voice heard in a shot, it is the voice that the viewer will be attentive to over the other sounds.[14] While The Brother's mouth is not in close-up (The Brother is in a medium shot), a viewer is likely to be attentive to his mouth since his mouth is part of his face and his mouth is wide open when he is crying out into the telephone. Chion argues that telephone conversations are an example of *on-the-air* sounds since they 'are not subjected to "natural" mechanical laws of sound propagation'.[15] Because The Brother is crying out towards the direction of the cinema screen, it is the 'air' in front of The Brother's face that is most likely going to be the focus of a cinema viewer's attention, an 'air' that in this shot is filled with the vibrations of his cry. The Brother's cry might be unheard, by both the sisters and the viewer, but nonetheless the force of The Brother's cry is so great that for the viewer the 'air' momentarily puts The Brother's face out of focus, even though there has been no actual shift in the camera's lens.[16] Across these shots from *An Unseen Enemy*, The Brother's mouth has become a source for a sonic force from within the film itself, a sonic force that is an affect since it is not actualized in the film, or for the cinema viewer, as an audible sound.

Henri Bergson in *Matter and Memory* offers a model for understanding how something that is not fully actualized might take the form of an affect. Bergson argues that the nervous system of a body is an intermediary system of sensation that exists between the reception of excitations and a responsive action. Bergson calls the nervous vibrations that form the suspensions or tensions existing between reception and reaction 'affections'.[17] At the beginning of *Matter and Memory*, Bergson describes affections as being 'an invitation to act, with at the same time leave to wait and even to do nothing'.[18] From his perspective, affections of the nervous system are a type of nervous tension – or suspension – that exists between excitation and response. This nervous tension, these affective states,[19] or affects, are therefore the potential actualization of a response, an action, that has not taken place . . . as yet. Here, we might understand the cry of The Brother in *An Unseen Enemy* as a force that has not yet formed into an audible sound. My approach to an inaudible sound is influenced by Eugenie Brinkema's argument that *near inaudibility* 'is a form of offering, of arrangement, of tension in a presentation that is always yet to take place'.[20] Similar to what Brinkema argues in regard to near inaudibility, that there is a 'tension . . . that is always yet to take place', The Brother's cry in *An Unseen Enemy* is a sound that is actualizing rather than actualized. The vibrations that are projected from The Brother's mouth are not fully actualized as a sound but are instead a vibrational aggregate that constitutes the becoming 'form' of inaudible tremors.

In the case of *An Unseen Enemy* the lack of actualization of a sound means that The Brother's cry is not heard by either the other characters in the diegesis or by the viewer. In her essay, Brinkema distinguishes silence from near inaudibility. She argues that while silence is generally figured and theorized as an absence of sound, that near inaudibility instead: 'advances pressure and intensity and force in place of presence and absence, plenitude and nothingness, produces a relationship to force based on form'.[21] The

Brother's cry in *An Unseen Enemy* is inaudible, but it nonetheless constitutes a pressure, an intensity and a force, which is felt as propulsion, a projection – a propulsion and projection that destabilizes the viewer's relationship to the visuals that are projected onto the cinema screen that is also a type of surface. To understand the type of surface that cinema screen is, we can also look at how Bergson describes surfaces in *Matter and Memory*.

In *Matter and Memory*, Bergson describes how the surface of what he calls an *organ of sense* is an intermediary that exists between the outside and inside of the body. Bergson argues: 'This organ is constructed precisely with a view to allowing a plurality of simultaneous excitants to impress it in a certain order and in a certain way, by distributing themselves, all at one time, over selected portions of its surface.'[22] Bergson argues that these excitations that impress the surface are an operation comparable to 'an immense keyboard'.[23] A surface for Bergson is like a keyboard and earlier in *Matter and Memory* in relation to sensation he argues that a surface is 'the common limit of the external and internal, is the only portion of space which is both perceived and felt'.[24] The cinema screen can also be understood as an organ of sense as it is described by Bergson, as a cinema screen is a 'common limit of the external and internal', of the cinema and of the film's diegesis and its mise en scène. The cinema screen can also be likened to a keyboard where the pressing of the keys takes place from both sides of the cinema screen, both from outside the film (the cinema) and from inside the film (the film's story world). For Bergson what is impressed on the surface of an organ of sense can be both perceived and felt. In *An Unseen Enemy*, The Brother's cry is felt as a force and is perceived as a cry, but it is not heard as a sound. His cry creates an impression on the cinema screen from within the film and in doing so brings attention to how the cinema screen is a limit between the external (cinema) and internal (film's diegesis): it is at this limit, within the intermediary system of the cinema screen, that sound affects manifest.

Tremors and tornados

In Griffith's *Broken Blossoms*, there is a sequence where the film's protagonist, Lucy Burrows, played by Lillian Gish is seen to scream in fear for her life. Unlike the scene in *An Unseen Enemy*, this scene is played as a series of shots of Lucy's face, shots that are in the same tight diegetic space – the cupboard, but which are nonetheless separated by four edits (the edits of this sequence are an example of an early form of jump cut). In *An Unseen Enemy*, two orphans are locked in a room, differently in *Broken Blossoms* Lucy locks herself in a cupboard to escape from her rampaging, axe-wielding father. In

The Sealed Room a wall is built to enclose two people having an affair, a wall
that constitutes the edge of frame. In *Broken Blossoms* the cupboard is set
within the wall itself. In the cupboard sequence in *Broken Blossoms*, it is as if
the spatial hole that separated and linked the two rooms in *An Unseen Enemy*
has become the space in which Lucy has to hide from the murderous attacks
of her father (Donald Crisp). What was once the edge of the frame in Griffith's
earlier two films become a frame that is actively bulging. Instead of Griffith
using parallel montage techniques and cutting in between different spaces as
it is in *An Unseen Enemy*, in the cupboard sequence of *Broken Blossoms* four
shots are cut within the same space. It is not the spatial difference between
shots that becomes key; rather, it is the way that these four shots appear
to echo each other within a single frame. Rather than an echo generating
difference through repetition of the same word that is heard after a gap of
time, the echo produced by the jump cuts in this sequence dislocates the
image from within. As the film historian Paul O'Dell in *Griffith and the Rise
of Hollywood* argues, the cupboard sequence from *Broken Blossoms* 'is one of
the rare moments in silent film history where the power of the image attains
an almost aural perspective'.[25]

   In relation to a different film directed by Griffith, *A Woman Scorned*
(1911) Tom Gunning describes how an empty space can become emphasized
through the contrast of two different shots. He says:

> A doctor (Wilfred Lucas) has been lured to an apartment by a gang of
> thieves who knock him unconscious. As he falls, we cut from this action
> to his wife and child at home sitting down to supper. In this shot, the
> father's empty place at the table is prominent in the left foreground. . . .
> [I]t is worth pointing out that this empty place occupies the same area of
> the frame that Lucas collapsed into in the previous shot.[26]

The two shots in *A Woman Scorned* invert the type of image relationship that
can be commonly found on a thaumatrope. Thaumatropes are an optical toy
that commonly opposed images, such as that of a bird and an empty cage.
When, for example, the thaumatrope of the bird and the cage is spun, the act
of spinning makes it appear as if the bird was imprisoned in the cage.[27] In
contrast to this Thaumatrope, *A Woman Scorned* focuses on the empty space
that is left by the bird once it has entered the cage, an empty space that is also
emphasized in a subsequent shot because the empty space occupies the same
part of the shot in each. In *Broken Blossoms*, the cutting up of a single space
makes it appear that Lucy is imprisoned in the cage and then in the cage of
the cage again, and again, and again; she is thus no longer one of the images
of the Thaumatrope but is arguably the act of spinning itself, an analogy that

is demonstrated in each shot as Lucy is physically spinning on the spot. In the four shots of the sequence that cuts from a medium shot, to another medium shot, to a close-up, to another medium shot, Lucy is seen to be crying out and screaming and spinning around; the act of spinning appears to be an attempt to physically amplify her cry beyond the confines of the closest that confines her.

In the cupboard scene of *Broken Blossom*'s Lucy scream is literally beside itself, becoming an affect that operates and is intensified across the four shots. The most intensive part of the sequence is when Lucy's face is in close-up where her screams and cries become a hyper-amplified expression that often literally puts her face out of focus (see Figure 13.3, Figure 13.4 and Figure 13.5). Lucy's scream in the close-up of her face can be understood as operating as the type of scream that Deleuze identifies in Francis Bacon's

Figure 13.3 Lucy Burrows (Lillian Gish) screams in the direction of the viewer in D. W. Griffith's *Broken Blossoms* (D. W. Griffith Productions, 1919).

Figure 13.4 Lucy Burrows (Lillian Gish) screams to the left in D. W. Griffith's *Broken Blossoms* (D. W. Griffith Productions, 1919).

Figure 13.5 Lucy Burrows (Lillian Gish) screams to the right in D. W. Griffith's *Broken Blossoms* (D. W. Griffith Productions, 1919).

paintings. For Deleuze, the painted screams in Bacon's paintings constitute the means by which the body in them tries to escape.[28] Across the jump cuts in this sequence it is Lucy's scream at death, her scream at being silenced, that is being amplified as if she was trying to escape through her own amplifying voice. In *An Unseen Enemy*, the hole in wall is both the threat of death and an unlikely means of escape. In *Broken Blossoms*, Lucy attempts to escape the confines of the cupboard she is imprisoned in, and through, her cries and screams. It is these cries and screams that vibrate in, and through, the walls that confine her, including that of the cinema screen that is acting as the cupboard's fourth wall.

As we saw, in *An Unseen Enemy*, the 'air' that is in front of The Brother's face is a type of projection that has been muted because it has nowhere to go. The shot of The Brother crying out is an example of a voice that has been silenced, a silenced voice that masks the face as the moment that The Brother is cut off from his sister. In *Broken Blossoms* the four repeated shots of Lucy in the same space create a synthesized scream that as an expression appears to be amplified above the four spatially incongruous shots (the cutting up of the space is what makes this space incongruous). In *Broken Blossoms* the four frames of the shot, which also constitute the walls and door of her cupboard, restrict the potential of Lucy's cries and screams to be heard in the film's diegesis by anyone else but her murderous father.

In *An Unseen Enemy* a gun protrudes through a hole in the wall to silence the two women, and, in *Broken Blossoms* Lucy is in the hole in the wall and is silenced. In the shot with Lucy's face out of focus, her screaming fills the space that is in between her and the viewer: the screams fill the space as there is nowhere else for them to go (like a room without windows that is filled with smoke). The momentary cry of The Brother in *An Unseen Enemy* becomes an operation of a whole scene in *Broken Blossoms* that takes place as if it were made of multiple close-ups of her scream and cries. It is as if these multiple close-ups were attempting to create a hole in the screen for Lucy to escape through. Balázs argues that the close-up reveals how movement is composed of smaller particles, much like the grains of dirt in a landslide.[29] In the shots from *Broken Blossoms* that I have been discussing, the smaller particles of movement of her cries and screams form an ineffectual spinning tornado, one that is unable to carry Lucy to safety. It is the tornado that is composed of tremors forming an earthquake – an earthquake as tornado and tornado as earthquake.

Similar to how the narrator's victim in 'The Cask of Amontillado' is silenced, in these three Griffith films, protagonists are threatened with being silenced (*An Unseen Enemy*) or are silenced (*The Sealed Room*, *Broken Blossoms*). In each of these three films, before the protagonists are silenced, in some cases permanently silenced, the inaudible tremors of the cry and screams manifest even if they are never fully actualized as a sound. By being attentive to these inaudible tremors, we have been attentive to voices that are violently silenced. We have also learnt something about how Griffith adapted Poe's 'The Cask of Amontillado' for the cinema.

Outside the context of Griffith's films, what does identifying and conceptualizing inaudible tremors as a sound affect enable us to do? I argue that non-actualized sounds – sound affects – can operate as signals for voices that cannot otherwise be heard. If, as Charles Sanders Peirce argues, indexical signs point to other things and force us to give 'attention to the particular object intended without describing it',[30] inaudible tremors are a signal for that which is no longer available to be pointed to or that which will no longer be available to be pointed to. Different to how Peirce conceptualizes the index, inaudible tremors need to be described in order for us to be attentive to them. One of the examples of the index that Peirce provides is that of the footprint.[31] Inaudible tremors can be considered as the becoming footprint of a sound or sounds that have not been actualized since these sounds have been prematurely silenced.

As a sound affect, inaudible tremors draw attention to that which cannot be heard, to that which is being silenced in, and on, and through, our relationship with cinema screens. Inaudible tremors cannot be seen, and they cannot be heard. However, inaudible tremors are signals for voices that are threatened to be, or that have been, silenced. It is because of this threat of silencing, or actual silencing, that these voices are unable to be actualized in the scene that they otherwise feature in. Perhaps then in other films, voices already silenced in a scene could be identified if we look for, listen out for, or attempt to feel out, the inaudible tremors that vibrate in and through and across our relationships with cinema screens? Perhaps we could also start identifying invisible and inaudible 'footprints' of those who have already disappeared or been erased or been silenced in films? While such questions are for future investigation for now I can say the following: it is by being attentive to inaudible tremors in a film that we can identify, be witnesses to, and be agitators for: 'Those who are screaming but cannot be heard; those who are bricked into cells with no release and those who cannot speak of their suffering.'[32]

I would like to thank Jodi Brooks and Sarah Cooper for their critical insights and editorial suggestions, when parts of this entry were being used for a different purpose. I would also like to thank Colin Campbell Robinson and George Damalas for their editorial suggestions on this entry.

Notes

1 Also known as *Broken Blossoms or The Yellow Man* [*sic*] *and the Girl.*
2 Edgar Allan Poe, 'The Cask of Amontillado', in *The Fall of the House of Usher and Other Writings: Poems, Tales, Essays and Reviews*, ed. with an Introduction and notes by David Galloway (1846; London, New York, Camberwell, Toronto, New Delhi, Auckland and Rosebank: Penguin Books, 2003), 315.
3 See Eugenie Brinkema, *The Forms of the Affects* (Durham, NC and London: Duke University Press, 2014).
4 Gilles Deleuze, *Cinema 2: The Time-Image*, trans. Hugh Tomlinson and Robert Galeta (1985; Minneapolis: University of Minnesota Press, 1989), 280.
5 I use capital 'T' for 'The' when 'The' is part of the character's name as given in the film's credits.
6 Jacques Aumont, 'Griffith – the Frame, the Figure', in *Early Cinema: Space, Frame, Narrative*, ed. Thomas Elsaesser with Adam Barker (London: BFI Publishing, 1990), 354.
7 Aumont, 'Griffith', 354.
8 Michel Chion, *The Voice in Cinema*, trans. and ed. Claudia Gorbman (1982; New York: Columbia University Press, 1999), 8.
9 Béla Balázs, *Theory of the Film: Character and Growth of a New Art*, trans. Edith Bone (1952; New York: Dover, 1970), 61.
10 See Rick Altman, 'Moving Lips: Cinema as Ventriloquism', *Yale French Studies* 60 Cinema/Sound (1980): 74.
11 The character is referred to as 'his favoured one' in an intertitle and the credits.
12 Aumont, 'Griffith', 354.
13 See Tom Gunning, 'Weaving a Narrative: Style and Economic Background in Griffith's Biograph Films', in *Early Cinema: Space, Frame, Narrative*,

ed. Thomas Elsaesser with Adam Barker (London: BFI Publishing, 1981), 340–47.

14 Chion, *The Voice in Cinema*, 6.

15 Michel Chion, *Audio-Vision: Sound on Screen*, trans. Claudia Gorbman with a foreword by Walter Murch (1990; New York: Columbia University Press, 1994), 76.

16 For a similar approach to breath in the cinema, see Davina Quinlivan, *The Place of Breath in Cinema* (Edinburgh: Edinburgh University Press, 2012).

17 Henri Bergson, *Matter and Memory*, trans. Nancy Margaret Paul and W. Scott Palmer (1896; New York: Zone Books, 1991), 17.

18 Bergson, *Matter and Memory*, 17–18.

19 Bergson, *Matter and Memory*, 53.

20 Bergson, *Matter and Memory*, 229.

21 Eugenie Brinkema, with composition by Evan Johnson, 'Critique of Silence', *Differences: A Journal of Feminist Cultural Studies* 22, nos 2–3 (2011): 223.

22 Brinkema, 'Critique of Silence', 128.

23 Brinkema, 'Critique of Silence', 128.

24 Brinkema, 'Critique of Silence', 57.

25 Paul O'Dell, *Griffith and the Rise of Hollywood*, with assistance of Antony Slide (New York: A. S. Barnes & Co; London: London: A. Zwemmer Ltd, 1970), 124–5.

26 Gunning, 'Weaving a Narrative', 342.

27 Jonathan Crary, *Techniques of the Observer: On Vision and Modernity in the 19th Century* (1990; Cambridge, MA and London: The MIT Press, 1992), 105.

28 Gilles Deleuze, *Francis Bacon: The Logic of Sensation*, trans. with an introduction by Daniel W. Smith and an afterword by Tom Conley (1981; Minneapolis: University of Minnesota Press, 2003), 24.

29 Balázs, *Theory of the Film*, 55.

30 Charles Sanders Peirce, *Peirce on Signs: Writings on Semiotic by Charles Sanders Peirce* (Chapel Hill and London: University of North Carolina Press, 1991), 181.

31 Peirce, *Peirce on Signs*, 252.

32 Colin Campbell Robinson on his reading of my essay and its contemporary relevance for identifying inaudible sounds in, and beyond, film.

Entry 14

Distortion

Greg Hainge

Crescendo

Distortion can be described as the alteration of the form of something, be this a sound wave or a solid object. Distortion then bridges the realms of the immaterial and material or, rather, conjoins them, demonstrating that no form is immutable, self-sufficient and self-identical across time. Distortion thus articulates us forcefully to a contemporary current in philosophy that elaborates an ontology premised on movement[1] as a means to escape the limitations of what Timothy Morton has termed philosophy's 'kinephobia'.[2] At stake here is the distinction between being and appearance, this being the very question around which any consideration of affect must turn. As Gregory J. Seigworth and Melissa Gregg write:

> Affect is integral to a body's perceptual *becoming* (always becoming otherwise, however subtly, than what it already is), pulled beyond its seeming surface – boundedness by way of its relation to, indeed its composition through, the forces of encounter. With affect, a body is as much outside itself as in itself – webbed in its relations – until ultimately such firm distinctions cease to matter.[3]

Thinking through this understanding of affect in the sonic realm, we will begin, as does Eugenie Brinkema,[4] in the realm of silence, and follow her in leaving this behind to consider near inaudibility, but then leave her behind and go further towards overdrive and distortion, by which time we will need to radically reformulate our opening line. This will be necessary because distortion will reveal to us how sound, when understood as an affect, can have no form, cannot be sound (in either the nominal or adjectival sense of this word) – and in doing so it will speak to us of much more than just sound.

From silence to near inaudibility

In her 'Critique of Silence', Brinkema takes issue with the dialectic often posited in discussions of sound (or noise) and silence, where the latter would constitute the (groundless) ground from which any sonic expression must necessarily arise. Posited as the a priori condition for there to be something other than silence – a scenario which brings about the annihilation of silence in the appearance of something other than silence and thus the eradication of the very ground or condition upon which the possibility of sound (or expression more generally) is founded – the realm of silence, for Brinkema, 'encompasses the radical impossibility of silence'.[5] Unpacking this further, she writes: 'Silence, perhaps, can have no relation at all from within a regime that posits it as the a priori, as what *is*, in the beginning. That it is the figure of silence that should be deployed for this metaphysical bind is appropriate: everything and nothing, it constitutes the fecundity and simultaneous annihilation of the sonic.'[6]

Pushing back against this radical impossibility upon which so much discourse in sound studies has been founded, Brinkema argues for a different set of concepts and axiomatics with which to think the sonic, and she finds these in the realm of near inaudibility:

> while the regime of silence is linked to a dialectic of being and nothingness, plenitude and finitude, the fullness of meaning or the ground of nonmeaning from which meaning is drawn, by contrast, a form that commits to near inaudibility is linked to a separate set of conceptual and aesthetic terms: pressure, tension, intensity, and force. Silence and near inaudibility pose entirely different formal problems and set in motion opposing formal gestures; the former concretizes the discourse of silence into a concern with being, while the latter concerns itself with formal gradations of intensities and with duration.[7]

While agreeing that the conceptual and aesthetic terms Brinkema proffers here are indeed more appropriate for thinking the sonic, there remains in her formulation a new series of double binds and radical impossibilities. To begin with, on the side of inaudibility Brinkema retains the notion of *a form*, yet to follow through the logic of her argument is to understand that the intensive, kinetic (and affective) realm described by these conceptual terms is one in which there can be no such thing – a problem she attempts to avoid in her book *The Forms of the Affects* via pluralization.[8] Conversely (although it is related to this last aporia), Brinkema draws a distinction here between the realm of silence, which is concerned with being, and the realm of near inaudibility, which is interested only in 'formal gradations of intensities and with duration'.[9] This, however, is a false dichotomy, for the kinds of formal analyses via which Brinkema proceeds are so important precisely because they help us grapple with ontological questions.

That this false dichotomy is posited should perhaps not surprise us, for it repeats the dominant schema of Western philosophy in which movement cannot be accorded any kind of ontological status – what Morton describes as Western philosophy's *kinephobia*.[10] This is a phenomenon that has been systematically traced – and then refuted by the formulation of an ontology premised on movement – in a major work by Thomas Nail, *Being and Motion*. Nail's stated aim is to offer 'nothing less than a wholesale redefinition of all the major categories of traditional ontological inquiry – space, eternity, force, time, quality, quantity, relation, and so on – as well as ontological practice itself, from the perspective of motion'.[11] While the primacy of motion does require Nail to posit that 'historical being must have at least some minimal kinetic attributes', since without these 'we risk positing the miraculous ex nihilo origin of motion',[12] the ontology he proposes is neither absolute nor universal but, rather, what he terms a 'regional realist ontology' that is at one and the same time an ontological description and a reflection on the inscriptive and performative practices of ontological description (within a Western philosophical tradition) at different historical periods.[13]

If the need for a new form of ontological description premised on the primacy of motion presses upon us now as never before, Nail contends, this is the case for a number of reasons. First, our reality itself is characterized by all kinds of accelerated global flows and movements and is thus best described as liquid rather than solid, he suggests[14] – following in the footsteps of others who have characterized our contemporary modernity as 'liquid' or 'mobile', such as Zygmunt Bauman, Marc Augé, Manuel Castells, Paul Virilio and others.[15] Second, Nail writes, 'we have now entered a new historical-aesthetic regime . . . the age of the image',[16] an age in which the mobility, ubiquity, availability and transmissibility of the digital image requires a new mode of engagement since 'aesthetics can no longer be understood by the old paradigm of representation'.[17] Third, because if we do not understand motion as 'something that is fundamentally constitutive of beings themselves . . . we will fail to understand the most important scientific phenomena of our time, such as nonlocality, entanglement, tunnelling, and quantum gravity'.[18] In brief, 'If being is defined by the historical primacy of motion today, yet existing ontologies are not, then we need a new ontology'.[19]

In spite of his insistence on the historically situated nature of the ontology he proposes, Nail's work inevitably harbours at its core an implicit critique of prior ontological formulations that, while understandable in the context of the philosophical currents and inscriptive practices of their time, are nonetheless here served a corrective – itself necessarily subject to the potential need for a further future corrective to be issued. Indeed, Nail's critique here is aimed at the very same kind of problematic logic that Brinkema kicks against in her 'Critique of Silence' where (as we have seen) the metaphysical dialectic of sound and silence or being and nothingness brings into the fray a self-defeating, radical impossibility. Similarly, having asserted that being must necessarily be apprehended not in itself, as being qua being, but as processual, relational, mobile, constituted only of flows and folds that are distributed in a field, Nail writes:

> As continuous movements, flows are by definition infinite; that is, they have no discrete beginning point and no discrete endpoint, since these points would render their continuity discontinuous, at a point. Since nothing comes from nothing, being has no kinetic first or final cause. Flows are therefore neither created nor destroyed, only rearranged. If flows are infinite in motion, and being is composed of flows, then the movement of being must be infinite. If being had an end after which it was destroyed, we would have already reached it. If there was nothing before it began, it could not begin from nothing. Therefore, being has no beginning and no end – just like its flows.[20]

As Brinkema moves away from silence – a concept that brings into the fray the same 'internally contradictory hypothesis of ex nihilo creation: something from nothing' that we find in Nail[21] – and towards near inaudibility, we hear in her formulations a forceful resonance with Nail's idea of kinetic being as infinite. She writes: 'What is revealed in the tension of the suspensions of near inaudibility is the affirming of the possibility of formal intensity – form's *There is* followed by no negation. Not denegation, but supplement. Not *There is no*, but *There remains sound.*'[22] As she approaches her final formulations, however, the idea returns of form as something that *is*, as something that comes preformed but placed under duress by intensive movements in play at its limits:

> Near inaudibility . . . affirms the duration of the form that it takes and puts under suspended pressure, but it also affirms the possibility as such of the generation of forms that persist. *That leap into what is suspended on the limit of form* affirms intensity and the vitality of sound, it makes present the vitality of the possible, suspended at the moment it neither arrives nor departs, but lingers, for a while.[23]

In the context of Nail's kinetic ontology with which Brinkema's work seems to share so much common ground to this point, the invocation of this kind of form whose coordinates and dimensions are known in advance but put under pressure by intensive forces external to that form is somewhat problematic. Indeed, the radical suggestion of Nail's work is that there can be no such thing as discrete form or, rather, that form only arises as a secondary effect of the periodicity of the in-folding of a flow. He writes:

> If being is fundamentally flow (continuous movement), then discreteness would simply be a relative or regional stability of that flow . . .
>   Here is the crux of the ontological problem of movement: Either we begin with it, or we never get it. This is a fundamental question for ontology. Either we begin with discrete and static being and have to say that real motion is an illusion, or we begin with flow and are able to explain stasis as relative or folded forms of movement.[24]

Nail's point here is significant, precisely because it is by following through the ramifications of his argument that we can understand the import of Brinkema's work, that we can avoid the false dichotomy according to which the regime of silence is concerned with being while near inaudibility can only concern itself 'with gradations of intensities and with duration' such that, in the final analysis, it is reduced simply to 'joy'.[25] Indeed, Nail's work enables us to understand that Brinkema's analysis of these intensities and durations is an analysis of a becoming-form, of the formal operations and relationality of her objects of study and that, as a result, the questions that it broaches *are* ontological.

While it would of course be possible to theorize near inaudibility in such a way as to avoid this invocation of form that pulls the rug from under the possibility of understanding intensity (and affect) as fundamentally ontological questions, if we think about the experience of near inaudibility it is perhaps easy to understand why it is that the spectre of form is so easy to invoke. Indeed, in the realm of the nearly inaudible we strain to *hear*, to make sound cohere in such a way as to be present to us, to take on a form, as though there were a form waiting to be filled that would itself be a condition of our ability to make it pass into the realm of the audible. (And lest this seem like an unfair attack on Brinkema's work, let me say that the seeming ineluctability of form also rears its head at times in work such as Erin Manning's *Relationscapes*, which places great emphasis on the primacy of movement and, for the most part, is careful to talk only of 'dynamic form'.)[26]

Rather than a realm in which there remains always the spectre of a form to be filled, a desire to make matter form, let us then turn to a form that is not a form, that is premised on the very undoing or annihilation of form not as something that exists and must be undone but, rather, as an ontogenetic process in which the undoing of form speaks to the impossibility of form, of the necessarily entropic and kinetic nature of the universe. To do this and try then to find a sonic figure more able to resonate in sympathy with Nail's ontological taxonomy of *flow, fold, field*, let us move from the realm of silence and the near inaudible to the very loud, the overdriven, to distortion.

fff

If we started off by suggesting that distortion can be described as the alteration of the form of something, it should by now be apparent why this needs to be reformulated. As per our critique of Brinkema, indeed, this formulation as it stands posits the existence of a form that would be altered by some external force or pressure, this being, if we are to follow Nail in his assertion of the primacy of motion as the basis of all ontology, no less impossible than the problematic dichotomy that considers sound to emerge from silence, something from nothing. We need, therefore, to figure distortion as the latent tendency of everything to move towards a state of increased entropy, as the process via which all that might appear to us as form (due to the periodicity of the in-folding of a flow when abstracted from the infinite flow of being) can only become ever more formless and thus reveal form to be only ever a secondary effect of a fundamental, processual, ontological condition.

For Catherine Malabou, the kind of form that we are gesturing towards here, that is premised on the annihilation of that form in the same instant that it comes into being, is governed by a process she terms 'destructive plasticity'. Most commonly associated with her neurologically inflected psychoanalytic work that seeks to arrive at an understanding of a different kind of brain plasticity that does not require there to remain in place, as per Freud's work, a primitive psychic stratum,[27] Malabou's concept of plasticity (itself formulated in response to the thought of Hegel) can be used to understand much more than psychic states – as we see in her own work. Malabou contrasts the concept of plasticity with that of elasticity, the latter describing the kind of form found in Brinkema's description of near inaudibility and our original description of the phenomenon of distortion, according to which a form undergoes a change but then returns to its former state once the force bringing about the alteration of that form is no longer operational.[28] For Malabou, plasticity proceeds otherwise; it is a very particular kind of forming that imbues the objects, concepts or phenomena that it constructs with an inherent and paradoxical instability that arises out of the dual axes

of the French terms *plastique* (plastic) and *plastic* (plastic explosives). As she
explains:

> *Plastique* is a synthetic material that can take on a range of properties
> and forms depending on the usage it is destined for. The word *plastique*
> also calls up the idea of a substitute, imitation or ersatz. And then there
> is *plastic* (plastic explosives), from which we get the terms *plastiquage*
> (a plastic explosives attack) and *plastiquer* (to blow up), which is an
> explosive substance made of nitro-glycerine and cellulose nitrate that
> can cause violent explosions.[29]

Combining both of these connotations, the plasticity we find in Malabou's
work – a concept that, as she writes, explains all that she has said about the
body of time and the body of the subject[30] – is 'at one and the same time
the emergence of a form . . . and the possibility of the annihilation of that
form. As if the taking and giving of form – *plastique* – were from the outset
contemporaneous to the explosion of form – *plastic*.'[31]

While still employing the notion of 'form', the kind of form proposed
by Malabou is then a form that is not a form, not something from nothing
but, rather, a form that is premised on the very annihilation of form or,
rather, form as an ontogenetic process that cannot exist in a preformed
state. Here we move towards the kinds of scientific phenomena that Nail's
ontology wishes to help us understand, for it is in such a manner that loop
quantum gravity theory conceptualizes time, which can never be imagined
to pre-exist a relation, is resolutely not a plane in which bodies, forms or
objects are situated but is, rather, generated out of the non-commutativity of
physical variables in relation with each other.[32] Malabou's plasticity is then
an operation according to which entropy arises from the very expression
of form – rather than something to which form is subjected in a universe
described as entropic, as is generally the case in common parlance and the
Second Law of Thermodynamics. The sound of this plasticity is distortion,
for distortion is a sound formed via an affective modality, existing only in the
in-between and thwarting the coalescence of form.

In the 1950s, distortion in the sonic realm was most often produced
through some kind of destructive act, be this deliberately increasing the gain
beyond the capacity of an amplifier or damaging the equipment in some way.
In both cases, the material assemblage of an amplification system is no longer
able to reproduce an audio signal without augmenting it with excessive
artefacts produced by its own materiality placed under duress, to the extent
that the signal's integrity, or (apparent) form, starts to come undone and to
express something of the assemblage out of which it is produced – that would

normally be sublimated under the purity of the signal. Think of a speaker
cone pushed beyond its capacity to the point where its structural integrity
is compromised and the resultant impact on the sound waves it produces
– an alteration of form from which there is no coming back, characterized
by plasticity not elasticity. The use of overdrive and the invention of effects
pedals would eventually enable musicians to use signal processing of various
kinds to achieve similar sounds without the need for such destruction, yet
distortion never really lost its destructive overtones, becoming the sound
of choice for bands born out of and kicking against the post-industrial
disaffection of their communities – like Black Sabbath – or entire genres
characterized by an aggressive rejection of societal norms and corporate
mass culture – such as punk.

Distortion thus provides us with a privileged site through which to
understand the ways that destruction can be generative of new forms of
expression and identity (and perhaps how we might map this concept onto
other systems and forms to destabilize them in turn). However, while the
very literal and exaggerated instances of distortion we are gesturing towards
here (and to which we will shortly turn our attention) exemplify in a very
unambiguous manner what is at play here, in the context of Nail's ontology
of motion our suggestion is that distortion is the operation produced by
plasticity, which, in turn, captures well the infinite, processual nature of
being conceptualized as a flow that is folded into semi-stable states that
are necessarily ephemeral. Distortion can then be figured as the expression
of the inherent plasticity of all being to point towards an ontological state
premised on 'the destruction and the very annihilation of all form' *from
within* or rather, going further still, on the negation of the very possibility
of form.[33] We find the term, indeed, in Gilles Deleuze's transcendental
empiricism where it undoes the fixity of sensory perception understood
as an act of recognition to highlight rather the forces produced in any
encounter. Deleuze writes:

> The point of sensory distortion is often to grasp intensity independently
> of extensity or prior to the qualities in which it is developed. A pedagogy
> of the senses, which forms an integral part of 'transcendentalism', is
> directed towards this aim. Pharmacodynamic experiences or physical
> experiences such as vertigo approach the same result: they reveal to us
> that difference in itself, that depth in itself or that intensity in itself at the
> original moment at which it is neither qualified nor extended.[34]

It is only in states such as this that affect can arise, following a double
negation that strips being of its illusory appearance of fixity and the

individual of any misguided sense of absolute autonomy or sovereignty. Here we find ourselves not in the realm of subjectivity but, rather, what Seigworth and Gregg call 'the midst of in-between-ness',[35] where there emerge not forms but rather '*a form of relation* as rhythm, a fold, a timing, a habit, a contour, or a shape [that] comes to mark the passages of intensities (whether dimming or accentuating) in body-to-body/world-body mutual imbrication'.[36]

Double negation

Low's 2018 album *Double Negative* unfolds according to the logics we have unpacked here. What seems to have struck critics most about the album at the time of its release is that its nearly every moment is distorted to the point that its sound seems to disintegrate from the inside, saturated with the 'cold, dark hiss of overloaded speaker'.[37] Here 'the noise no longer underpins the song – it is the song'.[38] 'It is an album with noise coming out of its wounds' that 'conjures the exact inverse of the sort of beautifying restoration work done on the soundtracks of vintage films to remove thumps, hums, and crackles'.[39] Or, as Stephen Deusner puts it, we are presented here with 'shuddering blooms of static in place of snares, blurred whorls of noise for bass, sounds that are violence itself' and make every song sound 'strained almost to its breaking point'.[40]

Working in collaboration with producer BJ Burton, these sounds and effects were not grafted on to a polished end product (as is often the case) but, rather, integral to the crafting of the album. As Rich Juzwiak explains:

> On the surface, *Double Negative* may appear to be a collection of songs that were composed and then dismantled, a sort of electronic-indie answer to prefab distressed jeans. . . . But apparently, the process was much more integrated than merely building up to break down – the band would show up with rough sketches of songs and then hammer them out with Burton. In the process, the line between performer and producer was scribbled out in static.[41]

As a result of this modus operandi, the songs on the album, as many reviewers have commented, sound like nothing ever heard before, like an entirely alien form of sound that has come from somewhere else, and this goes not only for their sonic material but their structure too. Deusner, for instance, writes:

The songs on *Double Negative* never move in expected ways: There's nothing that you'd really call a chorus or a bridge, just melodic passages that sometimes repeat and often do not. Songs fade out before the track ends, or they bleed into each other, or they just dissipate into the air. Disarticulated and discombobulating, the album never lets you get comfortable, never offers anything like solid ground.[42]

Or, as Ryan Leas puts it, '*Double Negative* seems to be intended for, seems to unfold, both everywhere and nowhere.'[43]

This impression no doubt comes, in part, from the way that different elements of the album – be these the songs, the instrumentation or any other aspect of it – operate according to '*a form of relation* as rhythm' – to pick up the formulation of Gregg and Seigworth once again.[44] Opening track 'Quorum', for instance, throbs for its duration, the metre of its rhythmic systolic and diastolic contractions determining that of the double time looped heartbeat percussion that marks the seamless transition point into the next track, 'Dancing and Blood'. As Deusner describes it:

'Quorum' opens the album with a blast of distortion, which coagulates into rhythmic, vertiginous waves. When Alan Sparhawk and Mimi Parker begin to sing, their voices are distorted and scrambled, yet their harmonies remain oddly intact, even catchy. Everything sounds distressed, and scarred, raw like an exposed nerve. As that song melts into 'Dancing & Blood', the commotion morphs into a strange, insistent beat, like a pounding heart or a timer counting down to some awful event.[45]

The awful event in question here is our contemporary reality, not the natural entropy of the universe but, rather, the accelerated self-destruction from within of our [read human] world. As Ben Beaumont-Thomas writes in his review of the album, subtitled 'the sound of the world unravelling': 'the erosion of America and our wider ecosystem, and the psychic state of living amid that erosion, is the focus here, enacted in the very music as well as the lyrics'.[46]

While some have wanted to find more hope in the album (Bill Pearis, for example, finds recurring themes that touch on 'the fight to keep going, to keep hope alive – even as their voices become lost in distortion and noise'[47]), the lyrics of closing track 'Disarray' seem to suggest that it may be too late for hope, that our inability or unwillingness to recalibrate our relation with the world and understand the extent of our 'world-body mutual imbrication'[48] has already brought about our undoing:

Before it falls into total disarray
You'll have to learn to live a different way
Too late to look back on apocryphal verse
And to be something beyond kinder than words
. . .
Not up for question, it's not even a thought
Another portrait you can hang on the wall
Dissolve into a state of awful inverse
The truth is not something that you have not heard.[49]

Alluding to the post-truth era of Trump, the double negative of this last phrase proved remarkably prescient, as the review in *Pitchfork* remarks.[50] This is not the only way in which we can read the album's title, however, for pictorially a double negative would be represented as two dashes, the start of a pulse, a rhythm. As per Nail's ontology and the rhythmic principle according to which this whole album is structured, a pulse is merely the periodic enfolding of an infinite flow, a pulse that, at the end of 'Disarray', does not stop dead but fades out, sounding, behind the scenes, into infinity. While the prognosis delivered here for our own human life forms may then be catastrophic, true to the operations of a kinetic ontology composed of infinite flows operating according to the principles of destructive plasticity, all this means in the final analysis is that new relations, rhythms and affects will come to replace them. As guitarist and vocalist Alan Sparhawk sings in *Double Negative*'s most bleakly optimistic line, 'It's not the end, it's just the end of hope.'[51]

Notes

1   See in particular Thomas Nail, *Being and Motion* (Oxford: Oxford University Press, 2019); see also Timothy Morton, 'Appearance is War', in *The War of Appearances: Transparency, Opacity, Radiance*, ed. Joke Brouwer, Lars Spuybroek and Sjoerd van Tuinen (Rotterdam: V2_Publishing, 2016), 166–82; Brian Massumi, *The Principle of Unrest: Activist Philosophy in the Expanded Field* (London: Open Humanities Press, 2017); Erin Manning, *Relationscapes: Movement, Art, Philosophy* (Cambridge, MA: The MIT Press).

2   Morton, 'Appearance Is War', 167.

3   Gregory J. Seigworth and Melissa Gregg, 'An Inventory of Shimmers', in *The Affect Theory Reader*, ed. Melissa Gregg and Gregory Seigworth (Durham, NC: Duke University Press, 2010), 3.

4 Eugenie Brinkema, with composition by Evan Johnson, 'Critique of Silence', *Differences: A Journal of Feminist Cultural Studies*, 22, nos 2–3 (2011): 211–34.
5 Brinkema, 'Critique of Silence', 211.
6 Brinkema, 'Critique of Silence', 216.
7 Brinkema, 'Critique of Silence', 213.
8 Eugenie Brinkema, *The Forms of the Affects* (Durham, NC and London: Duke University Press, 2014), xiii.
9 Brinkema, 'Critique of Silence', 213.
10 Morton, 'Appearance is War', 167.
11 Nail, *Being and Motion*, 24–5.
12 Nail, *Being and Motion*, 52.
13 Nail, *Being and Motion*, 33.
14 Nail, *Being and Motion*, 6.
15 Nail, *Being and Motion*, 38.
16 Nail, *Being and Motion*, 4.
17 Nail, *Being and Motion*, 4.
18 Nail, *Being and Motion*, 4.
19 Nail, *Being and Motion*, 16.
20 Nail, *Being and Motion*, 77.
21 Nail, *Being and Motion*, 73.
22 Brinkema, 'Critique of Silence', 231.
23 Brinkema, 'Critique of Silence', 231; my emphasis.
24 Nail, *Being and Motion*, 69–70.
25 Brinkema, 'Critique of Silence', 231.
26 Manning, *Relationscapes*, 6.
27 Catherine Malabou, *The New Wounded: From Neurosis to Brain Damage*, trans. Steven Miller (2007; New York: Fordham University Press 2012), 59.
28 It is interesting to note here that the concept of elasticity is very prominent in Manning's *Relationscapes*.
29 Catherine Malabou, 'La plasticité en souffrance', *Sociétés & Représentations* 20 (2005): 33; my translation.
30 Malabou, 'La plasticité en souffrance', 36.
31 Malabou, 'La plasticité en souffrance', 36; my translation.
32 See Carlo Rovelli, *The Order of Time* (London: Penguin Books, 2017), 122.
33 Catherine Malabou, *Plasticity at the Dusk of Writing: Dialectic, Destruction, Deconstruction*, trans. Carolyn Shread (2005; New York: Columbia University Press, 2010), 67.
34 Gilles Deleuze, *Difference and Repetition*, trans. Paul Patton (1968; London: Athlone Press, 1994), 237.
35 Seigworth and Gregg, 'An Inventory of Shimmers', 1.
36 Seigworth and Gregg, 'An Inventory of Shimmers', 13.
37 Andrzej Lukowski, 'Low, *Double Negative*', *Drowned in Sound*, 13 September 2018, https://drownedinsound.com/releases/20433/reviews/4152016 (accessed 1 March 2021).

38 Mike Goldsmith, 'Double Negative, Low', *Record Collector Magazine*, 7 August 2018, https://recordcollectormag.com/reviews/album/double-negative (accessed 1 March 2021).

39 Rich Juzwiak, 'Low, *Double Negative*', *Pitchfork*, 14 September 2018, https://pitchfork.com/reviews/albums/low-double-negative/ (accessed 1 March 2021).

40 Stephen Deusner, 'Low – *Double Negative*', *Uncut*, 9 October 2018, https://www.uncut.co.uk/reviews/low-double-negative-107640/ (accessed 1 March 2021).

41 Juzwiak, 'Low, *Double Negative*'.

42 Deusner, 'Low – *Double Negative*'.

43 Ryan Leas, 'Album of the Week: Low *Double Negative*', *Stereogum*, 11 September 2018, https://www.stereogum.com/2013136/low-double-negative-review/reviews/album-of-the-week/ (accessed 1 March 2021).

44 Seigworth and Gregg, 'An Inventory of Shimmers', 13.

45 Deusner, 'Low – *Double Negative*'.

46 Ben Beaumont-Thomas, 'Low: *Double Negative* review – the sound of the world unravelling', *The Guardian*, 14 September 2018, https://theguardian.com/music/2018/sep/14/low-double-negative-review-the-sound-of-the-world-unravelling (accessed 1 March 2021).

47 Bill Pearis, 'Low – *Double Negative*', *Brooklyn Vegan*, 21 December 2018, https://www.brooklynvegan.com/brooklynvegans-top-50-albums-of-2018/ (accessed 1 March 2021).

48 Seigworth and Gregg, 'An Inventory of Shimmers', 13.

49 Low, *Double Negative* (2018) [audio recording], USA: Sub Pop.

50 Juzwiak, 'Low, *Double Negative*'.

51 Low, *Double Negative*.

Entry 15

Feedback

Manuel 'Mandel' Cabrera Jr.

Acoustic feedback emerges within a zone of precarity: a situation in which the possibilities for sounding and listening, for act and affect, become fraught with Damoclean dangers. We're moving a microphone here and there near some speaker, curating the sonic space around us to suit the needs of the moment, when suddenly that threatening whine arises, and we know we're in the zone of precarity. The sonic situation emerges, not so much as an object of manipulation but rather as something to which we're vulnerable and thus must cater, lest the tiniest of movements take the subtlest of sounds and magnify them until they spiral into intolerable shrieks. Typically, then, we withdraw in a rush to escape the zone, trying to restore our sense of mastery. We swing the mic away from the speaker and hurriedly adjust settings as sound suddenly strikes us as a predator in hiding, waiting to leap out in an instant and stab at our ears.

The zone's material principle is the escalating cycle. Sound is received and transduced into electronic signals, then amplified and transduced back into sound, and run back through the chain again and again in a so-called reinjection cycle but one in which the signal intensifies with each of the many trips through the loop. The nature of this sort of cycle is highly variable. The mechanical and electromagnetic properties of the equipment, the placement of transducers and the resonant qualities of the surrounding environment all condition the zone's contours: which movements, sounds and settings will put us at risk of uncontrollable escalation, and which ones will take us closer to safety. It's uncertainty of these conditions that makes the zone precarious. They are nuanced, dynamic and difficult to discern, so that when we're in it we're often probing in ignorance.

For some, though, the zone has nevertheless had its seductions. What began as an irritating side effect of the high-powered microphones, pickups, amplifiers and loudspeakers that became commercially available in the mid-twentieth century has ended up being, for musicians of many different stripes, a seductive source of new sounds. By using those sounds, they've taken up manifold relations to the zone of precarity: taming the escalating cycle, toying with it, channelling its furious power, inhabiting it as curious explorers.

However, musicians have also sometimes used feedback as a medium in which to embrace precarity as such, of which feedback's zone of precarity serves as a synecdoche and expression. The vulnerability, the risk, the non-mastery entailed by being inhabitants of the world – these make us creatures of affect. This isn't simply true in the narrow sense that we're subjects of special, receptive experiences (e.g. perceptions and emotions) that arise through the world's impacts on us. Rather, it's true in a deeper sense, articulated by the composer Eliane Radigue in her reflections on the significance, not just of music, but of sound as such: that we are eddies in the flow of the world's becoming.[1]

Although feedback issues from machines we've engineered to expand the scope of our control over sound, it emerges as what precisely exceeds and flouts that control. The form in which it does so is as primal howls, whines and growls that put us in touch with, as Radigue puts it, the 'breath, pulsation, beating, murmur'[2] that constitute the enormous vibratory continuum of which the audible range represents only an infinitesimal sliver. This continuum isn't merely one aspect of the world, as Radigue describes it. Rather, it's the oscillation of becoming itself – the intricately structured dynamism that suffuses everything around us as well as ourselves: what she calls 'the immense vibrating symphony of the universe', the 'cacophony punctuating the deep ever-present rhythm of the breath, pulsations, beating'.[3] In feedback, we hear silence with the volume turned way up: aspects of the vibratory continuum that are normally undetectable by our ears – sounds too quiet to be heard, frequencies implicit in mechanical parts – magnified to audibility by the escalating cycle. Feedback thus confronts us with nature's feral riches – the exquisite complexity of becoming that always underlies the seemingly controllable, perceptible surface of things. In doing so, feedback discloses what Radigue has elegantly called, in the title of her remarkable essay, 'the mysterious power of the infinitesimal': a phrase that describes the inexhaustible dynamism of the world itself, which in feedback is dramatically taken out of hiding, arresting our ears in all its wild vibrancy.

Various artistic uses of feedback can, then, be seen as figures for our differing ways of responding to this dynamism, or the ways in which the world defies us and exceeds our grasp. Like Robinson Crusoe on the Island of Despair, some musicians enter the zone of precarity in a spirit of conquest, using the fine-grained techniques afforded us by audio technology and acoustic science to gain more and more meticulous understanding and control over what had previously seemed untamable. Others, however, have embraced non-mastery, that is, embraced affect. Among a number of other examples of this approach I will discuss below, Radigue's own use of feedback as an instrument provides us with some of the finest examples. That is, her remarkable feedback compositions are expressive of the sense that the world isn't something we confront in a struggle for control, but rather the perpetual flow of becoming that makes us what we are – makes us beings of affect through and through.

Most people probably first encounter musical feedback in rock and roll. We see footage of a guitarist soloing in notes that scream in ecstasy, playing as if the dance of their fingers over the fretboard and their waves of the neck are summoning occult forces, even expressing the musician's erotic potency (e.g. in the case of male guitarists for whom the instrument becomes positively phallic). We hear feedback used as musical punctuation: to create aleatory moments that add drive or weightlessness in between explosions of rhythm section and drums. At a concert, we witness it made to swell into an overwhelming wall of sound bombarding our whole bodies, not just our ears. Since the 1960s, when it entered the repertoire of superstars like the Beatles, the Who, Jimi Hendrix, Jefferson Airplane and the Velvet Underground, feedback has become a central part of the rock lexicon – a vehicle for its aspirations to incantatory power.[4]

However, during the 1960s, feedback was also taken up by artists with closer ties to conservatories, universities and the art world than to commercialized popular music. In the wake of developments like the rise of *musique concrète* and Karlheinz Stockhausen's bold experiments in using sound equipment, electronic and acousmatic music were in this period at the top of the musical avant-garde's agenda. As with rock, the increasing commercial availability of cheap, sophisticated electronics only intensified this tendency.[5] Musicians and composers laboured to craft themselves into acousticians and electrical engineers. They reconfigured sound equipment, scrabbled for spare parts and improvised idiosyncratic assemblages in search of new kinds of sound. Their efforts were often shaped by John Cage's turn away from music as meticulously crafted creation towards an appreciation of the endless treasures contained in sound as such – one made possible by the surrender of will and the embrace of accident.[6] Under this influence, many musicians and composers undertook experiments in search of sonic serendipity. It's no wonder, then, that many of them also became intrigued by the phenomenon of feedback.

Robert Ashley's *The Wolfman* (1964), one of the earliest feedback works to emerge from the avant-garde milieu of the 1960s, attempted to drag the audience kicking and screaming into the zone of precarity and deliver them over to feedback's predation. A single performer, clad in dark glasses and presented in the guise of 'a sinister nightclub vocalist, spotlight and all',[7] sings short phrases softly into a microphone as a tape piece assembled from processed found sound and electronic effects plays on the speaker system. However, the mic is fed into an amplifier cranked up to a level at which any sound causes feedback at incredibly high volumes, which is then modulated by the subtle changes in the shape of the performer's mouth as they vocalize, creating the impression that the feedback is swirling around the audience like a pack of screeching demons. This was years before rock music began its arms race towards increasingly rattling decibel levels and decades before artists like Merzbow and My Bloody Valentine cultivated reputations for live performances filled with deafening barrages of feedback and other noise. And so, Ashley's piece seemed sui generis, virtually unprecedented and became notorious.[8] He cast the vocalist in the role of the titular Wolfman, a murderous lycanthrope masquerading as the most innocuous of performers – a lounge singer. In doing so, Ashley had taken the most viscerally pleasurable of art forms, music, and daringly made it into an assault, a painful ordeal.

Other artists, though, worked to distance feedback from its image as nothing but a sonic predator. Ashley's piece had demonstrated a keen interest in transgressing and thus refiguring the space of performance and audition. And, the piece's presentation of its performer as its audience's aggressor suggested an implicit violence in the traditional division between audiences as those subjected to sound and artists as those doing the subjecting. *The Wolfman*, then, set the stage for works that would situate both audiences and performers as equally subject to the escalating cycle's vagaries – witnesses to, rather than manipulators of, its unfolding. By doing so, these works afforded new kinds of relations to the zone of precarity that the escalating cycle creates and thus to what is at issue in how it affects us.

Perhaps the most well-known example in this regard is Steve Reich's *Pendulum Music* (1968). The instructions for the piece dictate that several mics be suspended by cables from the ceiling, each above a speaker lying on its back to which that mic is connected through an amplifier turned up to the point of feedback. The piece's performers then pull back and release the mics so that they're swinging freely over the speakers, at which point the performers 'sit down to watch and listen like the rest of the audience'.[9] The result is a series of pulses, rhythmic and melodic but constantly changing according to the entropic nature of the process – the gradual slowing of the microphones' swings, which terminate with them at rest above the speakers, resulting in a hum of static feedback. Reich described *Pendulum Music* as 'the ultimate process piece' . . . 'audible sculpture. If it's done right, it's kind of funny'.[10] And its humour precisely comes from its narrative arc. As if giggling while goading a caged animal, *Pendulum Music* toys with feedback, repeatedly drawing its attacks but pulling just out of reach. Over time, though, it's as if this playfully reckless spirit is worn down by the predator's tirelessness and collapses into inert submission. And yet, when it does so, it enacts a moment of discovery. What, in evasion, had seemed a threat is, in yielding, unveiled as the serenity of a drone. In this way, the piece stages a gradual shift from the performers' initial manipulation of the zone of precarity, through the slow evaporation of this manipulation's traces by entropy, until a static vibration envelops performers and audience alike – all meditatively passive to the zone's workings.

In other sound pieces from this period, we find similar attempts to transfigure feedback, to see past its air of menace, and use it to connect us to the complex dynamics linking performer, equipment, audience and space. Among these, for example, are the works of Alvin Lucier.[11] As a co-founding member of the Sonic Arts Union, in the 1960s and 1970s, Lucier witnessed many innovative uses of feedback among his collaborators. This included Ashley's *The Wolfman*, which Lucier himself performed on 7 January 1967 at Brandeis University; as well as pieces like David Behrman's *Wave Train* (1966) and Gordon Mumma's *Hornpipe* (1967).[12] Such efforts were echoed in Lucier's own early work, much of which made unique use of reinjection cycles – that is, feedback, albeit not always of the sort that involves continuous signal escalation.

Lucier used reinjection cycles, for example, in his *Music for Solo Performer* (1965) in which, at John Cage's encouragement, he generated sound through biofeedback.[13] In this work, the performer is attached to electrodes that pick up their alpha brain waves. These signals are fed into loudspeakers and other transducers, whose agitations in turn activate percussion instruments. The resulting sounds are then heard by the performer, triggering brain waves that are fed back into the loop, so that their sensory and nervous systems become analogues for the microphone in acoustic feedback. Since alpha waves are autonomic responses – ones, in fact, requiring them to be as calm and still as possible – the performer is thus situated, not as the godlike creator of sounds, but simply as another element in the system, a participant on a par with the equipment and environment. Just as *Pendulum Music* stages the erasure of the boundary between performer and audience, *Music for Solo Performer* stages the erasure of boundary between performer and instrument. The performer becomes, not so much a creator of sounds, as one among many conduits through which the process unfolds.

In other pieces from this period, Lucier used a similar receptivity to achieve a kind of sonic palpation of the environment. For example, to perform his piece *Shelter* (1967), contact mics are placed on the walls, doorframes, and doors of an enclosed space (usually, a performance hall), allowing us to hear sounds outside it as filtered through the resonant properties of the architecture: for example, footsteps on the sidewalk, animal cries, musicians practising in far corners of the building – whatever lies outside the walls of the venue.

Through this process, we indeed experience the venue as a shelter, but not because its walls seal us hermetically in, fortifying us against a hostile world outside. Rather, it's because we're made to feel the world as precisely what bounds us within where we are, holding us, cradling us in place there. In such pieces, sound becomes a means, neither to observing things from a distance nor to pleasuring ourselves aesthetically through shaping it at will, but rather to arousing in our bodies and throughout our being the revelation of our co-presence with things and space.[14]

This latter theme converged with Lucier's use of reinjection cycles in his acousmatic piece *I am Sitting in a Room* (1969). To make its original version, he recorded a short text describing the process behind it, which begins:

> I am sitting in a room different from the one you are in now. I am recording the sound of my speaking voice and I am going to play it back into the room again and again until the resonant frequencies of the room reinforce themselves so that any semblance of my speech, with perhaps the exception of rhythm, is destroyed.[15]

As the piece progresses, Lucier's words grow ever more indistinct as they're repeatedly filtered through the room's material properties, recorded and reinjected in that recorded form back through the filter again. Gradually, it's as if the room itself begins speaking in its own voice of ethereal, undulating hums. Following Lucier's own description, the piece could be seen as the artist's erasure, accomplished through the disintegration of the speech by which he expresses his conception thereof. However, he goes on to describe the effect of this process as 'the natural resonant frequencies of the room articulated by speech'.[16] Destruction and articulation, then, are bound into one. But this is no willful paradox. Rather, Lucier's voice is treated here as simply part of a duet that unfolds through its individual voices – this speech, that vibration of walls – enchained in a dialogue of transduction. In this dialogue, each ceases to be what it was as it's transfigured through interchange with the other. But it isn't thereby destroyed, or subjugated by violence, but precisely enabled and amplified. A duet isn't a battle for domination, after all. As one voice in the duet, then, Lucier's own expression is made possible by the room's resonance just as much as it's made possible by him – the perfect encapsulation of a line from William Carlos Williams that Lucier has taken up as a kind of slogan for his work: 'No ideas but in things'.[17]

In later works, Lucier redeploys the theme of mutually transformative interchange with an environment to refigure the reinjection cycle in its escalating forms. For example, in *Bird and Person Dyning* (1975), his first major use of acoustic feedback as a primary element, a performer with binaural microphones attached to their ears wanders in the space immediately surrounding a tiny electronic birdcall – originally, a Christmas ornament he'd received as a gift.[18] The chirps are picked up by the microphones and played on loudspeakers, creating not only feedback but complex, spatialized forms that result from its heterodyning with the birdcalls. Lucier himself recounts his discovery of this effect: 'I began hearing phantom images of the birdcall, which seemed to come from inside my head and at the same time to be located in various parts of the room.'[19] Here, the very body of the performer, and not just their voice and their ears, is inserted into the zone of precarity. Their resulting dialogue with the environment takes the form of their silent slow wandering on the stage, exploring but also shaping a sonically articulated space in which the distinction between outside and inside is blurred. Neither manipulators of feedback, nor victims of its violence, they become instead inhabitants of the escalating cycle, revealed as simply the interpenetration and co-determination of performer, objects and environment.

In these pieces from Reich and Lucier, we find no hint of feedback as a threat – one to be teased, deployed as a weapon, or appropriated as a force all one's own. Instead, it's figured in a different way altogether, for which there's no better image than Radigue's aforementioned 'the mysterious power of the infinitesimal.'[20] Radigue's own journeys with feedback took place in relative isolation. During the mid-1950s and again in the mid-1960s, she'd worked for both Pierre Schaeffer and Pierre Henry, assisting with their forays into *musique concrète*.[21] She envisioned an entirely different kind of music of her own, though: a music of developments so slow as to be imperceptible, like the growing of plants.[22] However, neither Schaeffer, Henry, nor the coterie that surrounded them took her seriously as a musician. As she says: 'I was the little intern who no one paid attention to, especially in such a macho environment. . . . To demonstrate to you how much my presence was appreciated in that place: one day, a technician, who I won't name, said "It's great to have Eliane in the studio because she makes it smell good!"'[23] Thus, when Radigue took up feedback as her primary medium while working at Henry's Studio Apsome, and presented him with the result, *Jouet Electronique* (Electronic Toy) (1967), Henry dismissed it out of hand.

But she persisted anyway. After leaving Studio Apsome, Radigue found herself in the situation of being a single mother without access to a studio, left with only a few pieces of equipment Henry had bequeathed her: three tape recorders, a mixing board, an amplifier, two loudspeakers and a microphone.[24] Through meticulous, painstaking work, mostly done during the hours when her children were in school, she continued her tutelage in feedback, producing a series of acousmatic pieces assembled using acoustic and tape feedback, for example, $\Sigma = a = b = a + b$ (1969), *Usral* (1969), *Omnht* (1970), *Vice-Versa, etc. . .* (1970), *Stress-Osaka* (1970) and *Opus 17* (1970).[25]

In Radigue's feedback compositions, we find a sense for the cosmic flow of becoming, and of our embeddedness therein. Hums and chirps, rumbles and throbbing. We could easily mistake them for celestial energies as detected by satellites and radio telescopes: the inferno of stars, the electromagnetic songs of planets, or the thrum of the universe's microwave background. These are vibrations that aren't produced for consumption in a moment, that don't cater to the time of anxious bustle, but instead change at the pace of the tide. If we're distracted for a minute, we can suddenly realize the ambience has shifted without being able to pin down how or when: hints of a glacial history that stretches beyond, but includes, human affairs – the time of nature's unfolding over epic spans.

In this way, Radigue soars past Lucier's concern for our immediate environments into one for the world as a whole. Both cross into the zone of precarity and allow themselves to be swept up in the escalating cycle, precisely to get in touch with our embeddedness in things. In *Usral*, *Omnht* and *Stress-Osaka*, which were composed to serve as elements in larger sculpture or installation pieces, Radigue, like Lucier, demonstrated a keen sensitivity to the place of the sounds she produced using the escalating cycle in the situation of sounding and listening.[26] However, whereas Lucier's pieces are content to explore the spaces and materials of equipment, architecture, and bodies, Radigue's reach for the totality of what lies beyond all this: the 'immense vibrating symphony' that is the world itself.

To access the large, though, she looks to the small. With a view to the large, we find, she argues, 'the air's powerful breath, violent intimidating tornados, deep dark waves emerging in long pulsations from cracks in the earth, joined with shooting fire in a flaming crackling' – in short, all of the forces our subjection to which embodies our precarity; our vulnerable, dependent place in things.[27] Beneath them, though, there's the common thread: the undercurrent of vibration, transduction, wordless dialogue, what Radigue calls 'the ever-present rhythm of the breath, pulsations, beating' whose presence throughout us and in all our surroundings constitutes our intimacy with the cosmic.[28] We emerge from this undercurrent, our power a bit of its own – its power for interchange with itself: as she puts it, 'An organ adapts itself to transformation of a miniscule zone from the immense vibrating spectrum decoded into sounds captured, refined, meaningful', even as we begin our own 'crackling, roaring, howling and growling, the noises of life'.[29] From this power there issues our ordering of sounds in language and music: 'The subtle alchemy of sounds becomes, oh wonder, understood."[30] Yet, she immediately reminds us, 'breath, pulsations, and beating remain' as 'natural harmonics unfurl into space in their own language'.[31]

For Radigue, feedback is simply one form in which this language of 'breath, pulsations, and beating' speaks through us, yet therefore exceeds us: the murmuring of the infinitesimal foundations of being – enabling our expression, yes, but never, ever as a monologue: 'The occasional accident, a disrupted relation between recorder-transmitter-recorder-playback, and there our medium assumes some independence. How, then, does it behave? Breath, pulsation, beating, sustained sound, depending on the mood. So much richness in all this "feedback" and other chance or provoked "interference".'[32] To make contact with this murmuring, Radigue learned on her simple equipment how to play feedback, to let it play her. Endless hours over months and years, holding a microphone, patiently shifting it here and there by the tiniest of increments – a dance of near stillness whose moves were the subtlest of motions, her ears and hands letting themselves be choreographed by the most fragile of sounds. She entered into the zone of precarity, not to conquer it, toy with it or even just to explore it but to dwell there in contented equanimity, serving as a vehicle for the infinitesimal's untamed power.

In doing so, Radigue refigures the zone of precarity, and in the process the source and venue of precarity as such: the all-encompassing, ever-becoming world. The world becomes, not a space of danger, where anything that shapes us beyond our ken or control thereby enslaves and demeans us. Rather, it becomes the author of our being and the source of our power, with which we commune as much through accident as through deliberate effort. In apprenticing ourselves to it, we feel ourselves, not held hostage to it, but held in place by it, and our fragility as the conduit through which our own power is conferred: our power to embark upon 'adventures, explorations of this infinite mystery of the transmutation of noise into sound, of sound into music and, as with all true questions, to receive in response only a few "hows," never a "why," thus leaving endless freedom to trace one's path, to find one's voice'.[33]

Notes

1 Eliane Radigue, 'The Mysterious Power of the Infinitesimal', *Leonardo Music Journal* 19 (2009): 47–9.
2 Radigue, 'The Mysterious Power of the Infinitesimal', 47.

3 Radigue, 'The Mysterious Power of the Infinitesimal', 47–8.
4 For example, the Beatles's hit 'I Feel Fine' was released in 1964, before any
   of these other bands exploded into popularity. 'I Feel Fine' begins with a
   brief moment of guitar feedback that John Lennon once claimed was 'the
   first feedback ever recorded' that wasn't 'on some old blues record from the
   Twenties'. Cf. John Lennon and Yoko Ono, 'Interview with David Sheff',
   Playboy, January 1981, http://www.beatlesinterviews.org/dbjypb.int4.html
   (accessed 15 September 2021).
5 For a comprehensive history of electronic music, including of the impact
   of the commercial availability of sound equipment on its development, cf.
   Thom Holmes, *Electronic and Experimental Music: Technology, Music and
   Culture* (New York: Routledge, 2016), especially ch. 1, 'Early History: Prede-
   cessors and Pioneers (1874–1960)', 1–166. More specific but not restricted
   to the phenomenon of feedback, for a historical survey of the use of microphones and speakers in electronic music, cf. Cathy van Eck, *Between Air
   and Electricity: Microphones and Loudspeakers as Instruments* (New York:
   Bloomsbury Academic, 2018).
6 For a substantial selection of John Cage's own reflections on the abrogation
   of intention in musical practice, which he associates with the phenomenon
   of silence, cf. John Cage, *Silence: Lectures and Writings* (Middletown, CT:
   Wesleyan University Press, 1973).
7 Robert Ashley, 'The Wolfman for Amplified Voice and Tape', in *Source:
   Music of the Avant-Garde 1966–73*, ed. Larry Austin and Douglas Kahn
   (Berkeley: University of California Press, 2011), 143–5: cf. 145. The two tape
   pieces Ashley used for performances of *The Wolfman* were *The Wolfman
   Tape* (1964), and later *The 4th of July* (1962): 143.
8 In his highly influential history of experimental music, the composer and
   music theorist Michael Nyman calls *The Wolfman* 'possibly the best-known
   feedback piece', and its influence extended beyond avant-garde and into
   popular music. For example, seeing a performance of *The Wolfman* has been
   cited as one of Iggy Pop's main inspirations for the bold rock experimentation he undertook with his seminal band the Stooges. Cf. Michael Nyman,
   Experimental Music: Cage and Beyond (New York: Cambridge University
   Press, 1999), 100–1; and Benjamin Piekut, *Experimentalism Otherwise: The
   New York Avant-Garde and Its Limits* (Berkeley: University of California
   Press, 2011), 18 and 185–6.
9 Steve Reich, 'Pendulum Music', in *Writings on Music, 1965–2000*, ed. Paul
   Hillier (New York: Oxford University Press, 2002), 32.
10 Steve Reich, interview with Thom Holmes, *OHM: The Early Gurus of
   Electronic Music*, www.furious.com/perfect/ohm/reich.html (accessed
   12 February 2021).
11 I focus on Lucier only for reasons of space, and as one of the most well-
   known avant-garde composers who embraced non-mastery and coinci-
   dence in the use of reinjection cycles, and of feedback in particular. One

could easily approach similar themes through the work of other early feedback innovators. For example, there are Max Neuhaus's attempts at using feedback as a medium for enacting an immediate, non-linguistic form of community. Cf. his *Public Supply I* (1966) and *Max Feed* (1966), both outgrowths of his use of feedback in performances of John Cage's *Fontana Mix* (1958). Or, there's David Tudor's use of signal feedback in pieces like *Untitled* (1972), *Toneburst* (1975) and *Pulsers* (1976), which Tudor once described as entries in 'a never-ending series of discovered works in which electronic components are found to be natural objects'. Cf. David Tudor, *Pulsers/Untitled* (1984) [audio recording], USA: Lovely Music Ltd, liner notes.

12 Robert Ashley, 'Performances', *robertashley.org*, http://www.robertashley.org /performances/performance-history.htm (accessed 9 February 2021).

13 Lucier has recounted the importance of Cage's urging in the genesis of the piece, for whose first performance Cage served as an assistant, in a number of interviews and writings. For example, see Alvin Lucier, 'Alvin Lucier on a Lifetime of Experiments' [video], interviewer unlisted, Red Bull Music Academy, YouTube (uploaded 31 May 2017), https://www.youtube.com/ watch?v=v-Pnb_ZE7Hs (accessed 8 February 2021).

14 One could also mention, for example, Lucier's *Vespers* (1968), in which the performers are sealed in a pitch-black room and left to navigate it using the clicks of handheld echolocation devices.

15 Alvin Lucier, *I am Sitting in a Room* (1993) [audio recording], USA: Lovely Music Ltd.

16 Lucier, *I am Sitting in a Room*.

17 Taken from William Carlos Williams, *Paterson*, ed. Christopher MacGowan (New York: New Directions Publishing, 1995). The line appears in several places in the book, e.g. in the poem 'The Delineaments of the Giants', 16.

18 For Lucier's own account of the genesis of this piece, on which I've drawn, cf. Alvin Lucier, 'My Affairs with Feedback', *Resonance* 9, no. 2 (2002): 24–5.

19 Lucier, 'My Affairs with Feedback', 24.

20 This is the title of the aforementioned short but impressive essay by her on the nature of music: see note 1.

21 Emmanuel Holterbach, 'Eliane Radigue: Feedback Works 1969–1970', in the liner notes for Eliane Radigue, *Feedback Works 1969–70* (2014) [audio recording], Italy: Alga Marghen, 3.

22 Eliane Radigue, 'Interview with Eliane Radigue' [video], interview by Evelyne Gayou, trans. David Vaughn, YouTube (recorded 2013, uploaded 12 November 2018), https://www.youtube.com/watch?v=dByqwi7Jvbo&t =1088s (accessed 6 February 2021, 10:44–11:00).

23 Radigue, 'Interview with Eliane Radigue', 2:36–3:03.

24 Holterbach, 'Eliane Radigue', 3.

25 For her description of her work habits at this time, cf. Eliane Radigue, 'Interview about the Feedback Works: Paris, 5 April 2011', interview by

Emmanuel Holterbach, in the liner notes for Eliane Radigue, *Feedback Works 1969–70* (2014) [audio recording], Italy: Alga Marghen, 8. Strictly speaking, the opening passages of *Opus 17*, her final work from this period, do not exclusively employ feedback in the usual sense. Rather, Radigue used techniques that were, unbeknownst to her, nearly identical to Lucier's in *I am Sitting in a Room*, sending excerpts of Chopin and Wagner through reinjection loops until they transformed into shimmering walls of ambience. However, in the remainder of *Opus 17*, as well as all of the works leading up to it, acoustic and tape feedback were Radigue's only 'instruments'. Cf. Emmanuel Holterbach, 'Eliane Radigue: Opus 17', in the liner notes to Eliane Radigue, *Opus 17* (2015) [audio recording], Italy: Alga Marghen.

26 Cf. 'Interview about the Feedback Works', as well as 'Eliane Radigue: Feedback Works 1969–70', 3–5.
27 Radigue, 'Interview about the Feedback Works'.
28 Radigue, 'Interview about the Feedback Works', 48.
29 Radigue, 'Interview about the Feedback Works', 47–8.
30 Radigue, 'Interview about the Feedback Works', 48.
31 Radigue, 'Interview about the Feedback Works', 48.
32 Radigue, 'Interview about the Feedback Works', 48.
33 Radigue, 'Interview about the Feedback Works', 49.

Entry 16

Sub-bass

Aidan Delaney

There is a certain captivation with bass in contemporary consumer culture. This is evidenced in the tuning of curves of designer headphones that emphasize low frequencies and in much of the music often played through them. Bass frequencies induce movement, be it the vibrational modulation of inorganic bodies caused by resonant frequencies or organic bodies induced towards rhythmic motion. For the kinetic potential of sub-bass to breach the auditory threshold, it needs significantly more energy compared to higher registers. This is to say our perception of frequency of logarithmic and audible sub-bass is thousands of times more powerful than the range of frequencies found in speech.[1] Due to this power, the vibrational force of sub-bass can elicit a wide range of resonant affects, acting upon bodies to cause further vibrations and produce sensation. Contemporary popular music has increasingly exploited the affective quality of sub-bass for the listener's pleasure and perhaps more cynically to foster competitive advantages in capturing attention.[2] Steve Goodman asserts that within cultural studies bass demands theoretical attention, suggesting that its 'vibratory immersion provides the most conducive environment for movements of the body and movements of thought'.[3] This is the notion that Paul C. Jansen develops when he posits that grasping a plane of sonorous relations involves understanding 'bass-induced movements of bodily thought, which, in turn, means giving greater attention to the role of sensation in sonic culture'.[4]

However, while there are several positive affirmations of bass perception (through the likes of Goodman and Jansen), there is another side to this oscillating force. On the one hand, we might say that sub-bass evokes pleasurable resonance, on the other, we must acknowledge its capacity for corporeal vibrational aggravation. This situation is well documented in the field of noise studies where low-frequency sound is known for its potential to induce nausea and cause other ill effects.[5] It is this affective polarity of sub-bass that provokes Frances Morgan to write:

Bass has a physicality that is not shared by other registers: it is this that makes it comforting, immersive, pleasurable, as it mirrors natural rhythms and processes. It can also be alienating and dehumanising – low-end sounds can make you aware of your physical fragility, make you feel sickened, disorientated and crushed. Bass can also be profoundly sad, suggestive of vast, lonely space.[6]

Morgan's position here is caught between representation and sensation, but her point about the physicality of bass is worth returning to. In everyday life, the corporeal vibrational aggravation caused by sub-bass is often experienced as mechanical noise as a sonic industrial by-product or less often in nature as extreme weather or geo-activity. This is sub-bass and infrasonic noise as unwanted sound. While in contemporary music there is an increasing presence of extended low end, at times, intentionally composed to be physically overwhelming (think Dubstep), it is hard to approximate this as unwanted sound for its intended audience. Sub-bass in the cinema can play differently. Emphatic use of the low end to produce corporeal agitation is increasingly found in film as cinematic sound technology has developed. From the Sensurround technology first used in the film *Earthquake* (1974) to the object-based immersive sound produced by Dolby Atmos, moments of blockbuster spectacle are often intensified by pronounced sub-bass. Gaspar Noé's film *Irréversible* (2002) is a rather notable example where low-frequency sound is intentionally used to elicit affection in the audience. The film features an extensive and enduring sub-bass drone with fundamental frequencies modulating between 24 Hz and 35 Hz. Consequently, this film serves as a useful example for understanding sub-bass as an affect. Before attending to the affective aspects of *Irréversible*'s low end, it is beneficial to first acquaint the reader with its narrative and comment briefly on its cultural reception.

Irréversible can be succinctly referred to as a rape-revenge film, a subgenre of films that 'exemplify particular ideological treatments of sexual violence and its relationship to retribution'.[7] However, unlike the exploitation films normally associated with this subgenre, *Irréversible*'s narrative is revealed in reverse chronological order, meaning that the revenge antecedes the rape, cause precedes effect. Almost at mid-point in the film, Alex (Monica Bellucci) is brutally raped and beaten until comatose. However, the film begins with the reaction to this event, opening with her partner, Marcus (Vincent Cassel), and her ex-boyfriend, Pierre (Albert Dupontel), on a rampage to seek revenge on Alex's attacker Le Tenia (Jo Prestia). The film is composed of twelve narrative segments where the nocturnal violence retracts to the tender intimacies shared between the couple earlier that day. The story structure means that Alex's rape is foreshadowed long before we witness it. In the context of this onscreen rape *witness* is most appropriate. It is a nine-minute single take with no recourse to cinema's dissection of time and space, giving it a verisimilitude in the tradition of Bazinian realism. This is to say that there are no edits, and the once peripatetic camera seizes to move when the assault begins, leaving the audience to listen to Alex's pleas before her torture bears out in sonic brutality. In a word, *Irréversible* is a difficult film – both its form and content are challenging. Its form is composed of audiovisual devices – dizzying, serpentine-like camerawork, strobe lighting, over-saturated colours and ominous sound design – devised to viscerally provoke the audience. Such formal elements, as moments of narrative rupture and spectacle, can be deemed attractions: formal devices used to exercise 'a definite effect on the attention and emotions of the audience'.[8] Matt Bailey suggests that director Gaspar Noé 'consistently uses a systematic deployment of attractions outside of the narrative to elicit primal responses from the spectator', further adding that '*Irréversible* features a number of scenes that employ aural or visual amplifications or modifications in order to provoke an effect'.[9] Bailey here is suggesting that the film, and all of Noé's work, is a return to what Tom Gunning calls the 'cinema of attractions'. It is a type of cinema that is 'willing to rupture a self-enclosed fictional world for a chance to solicit the attention of the spectator'.[10] *Irréversible*'s content is equally provocative and it has been categorized within a cycle of French cinema associated with transgression, 'brutal intimacy' and a 'disturbing use of graphic physicality' with 'depictions of sexual and social disfunction', thus earning the appellation of *cinéma du corps* (cinema of the body).[11] The BBC described the film's reception at *Festival de Cannes* (Cannes Film Festival) in 2002 as 'so shocking that 250 people walked out, some needing medical attention', explaining that oxygen was administered to twenty people who fainted during the screening.[12] In conjunction with the extremity of the film's content, along with its dizzying visual formal elements, the soundtrack features overt sonic attractions that intentionally operate as vibrational forces to act upon its audience. Specifically, there is a deliberate weighting of the frequency spectrum towards the low end and excessive use of low-frequency noise to create affection. The concepts of *becoming* and the *materiality of sound* will be used to characterize this relationship between sub-bass and affect more generally and how it is exemplified in Noé's revenge-rape assault.

Christoph Cox argues that analysis and appreciation of sound art should employ a materialist approach, using what he calls sonic materialism. He contends that aesthetic theory has been dominated by frameworks centred on signification, representation and mediation. His project is to elevate sound discourse in the arts, arguing that sonic materialism offers a way to think about sound that accounts for its non-representational aspects. In examining the ontology of sound, he explains that while sources can '*generate* or *cause* sounds, sounds are not bound to their sources as properties. Sounds, then, are distinct individuals or particulars like objects.'[13] To illustrate this phenomenon Cox demonstrates that visual objects can survive alterations of their properties, but the properties themselves cease to linger. If a red door is painted blue the redness of the door does not remain. Sound is different. 'A sound that begins as a low rumble may become a high-pitched whine, while remaining a single sound.'[14] Cox further explains that sounds are not objects, but are temporal events and tied to the qualities they exhibit over time: 'If sounds are particulars or individuals, then, they are so not as static objects but as temporal events.'[15] This is sound understood as a relation, rather than an object – an important position for understanding sound as affect. The ontological, temporal aspect of the sonorous is critical to Cox's argument and he explicates this as a 'possibility of change' through an 'ontology of events', elucidated through the Nietzschean concept of 'becoming'.[16] This allows him to maintain that aesthetic theory should no longer relegate sound to theories of signification, asserting that sound is 'firmly rooted in the material world and the powers, forces, intensities, and becomings of which it is composed'.[17] While Cox's argument does not engage with affect directly, there are various indications towards it. He calls for artistic productions to be treated as material forces inflected by other forces: 'We might ask of an image or a text not what it *means* or *represents*, but what it *does*, how it *operates*, what changes it effectuates.'[18] This materialist, ontological framework, treating the sonorous as a possibility of change, is built on notions of intensity, event and becoming. These are key concepts in Gilles Deleuze's philosophy which can be put to work in thinking through the relationship between sound and affect, alongside sub-bass as a specific type of sound that produce affects, as evidenced in its use in *Irréversible*.

Throughout their collaborations Deleuze and Félix Guattari offer a productive ground for understanding affect. For them, affects are becomings. They explain: 'To every relation of movement and rest, speed and slowness grouping together an infinity of parts, there corresponds a degree of power. To the relations composing, decomposing, or modifying an individual there correspond intensities that affect it, augmenting or diminishing its power to act; these intensities come from external parts or from the individual's own parts.'[19] Consequently, the concepts of affect and becoming are co-constituted in Deleuzian thought. Affect is an interplay between bodies resulting in affection; becoming is a flux of transformation involving both movement and duration. Sound as affect is emblematic of this. Sound is composed of vibrating bodies acting on other bodies, and it transforms over time, therefore sound always has a temporal dimension. Intensity is also fundamental in understanding this relation. The concept is a key feature throughout Deleuze's writing to typify dynamics of difference and to characterize affect. Intensities can be 'a wave of vibrations, a continuous variation, like a terrible threat welling up inside us'.[20] There is a distinction made between force and intensity, 'force concerns "movement", while intensity concerns "speed" (in many ways, force can be considered in terms of that which puts series into communication, while intensity concerns the resulting difference)'.[21] This concept of intensity will be the ground by which sub-bass, as a unique sound affect, is later defined. In addition to this relation of affect and becoming through degrees of intensity, Deleuze conceives event as 'a vibration with an infinity of harmonics or submultiples', meaning that the event is never a singularity, it is a multiplicity of actualizations, where the future and the past of the event are contained within it.[22] From this we can conclude that *event* is *becoming*, it is 'that which has just happened and that which is about to happen, but never that which is happening'.[23] Therefore, sound as temporal flow is an intensity of affect that plays out as an event of material oscillations in a state of becoming.

In the context of *Irréversible*, it is important to note this relation of affect and becoming with respect to the sub-bass as a cinematic attraction where the low frequency is in an affective relation of intensity. The ideal listening experience of the film is through a cinema sound system, and this can be explored through the previous report of its premiere screening at the Cannes Film Festival. What can be said of a cinematic event where the film provokes a physiological affect so intense that its spectators require medical attention? The corporeal and sensational dimension of the film when screened in certain conditions has the potential for producing visceral affections in the audience through vibrational shock tactics in the sub-bass range. Analysing the film's frequency content using audio-metering software illustrates that the sound design, created by Thomas Bangalter, is dominated by low-frequency noise with low pass filtering to roll off some of the mid and high frequencies. This sound, as previously mentioned, is a sub-bass modulation with a fundamental frequency oscillating between 24 Hz and 35 Hz, and a range of higher-frequency harmonics and complex overtones (see Figure 16.1). In literature discussing the film, there are accounts of the fundamental frequency of the sub-bass pitched at a stable 27 Hz and its duration either a continuous thirty minutes or sixty minutes or throughout. For example, Matt Bailey notes that 'for the first sixty minutes of the film, [there is] a constant 27-hertz tone specifically designed by Noé to cause nausea in the audience'.[24] David Davis writes that 'the loud and already disorienting pulse of the soundtrack of the opening scenes of the film was

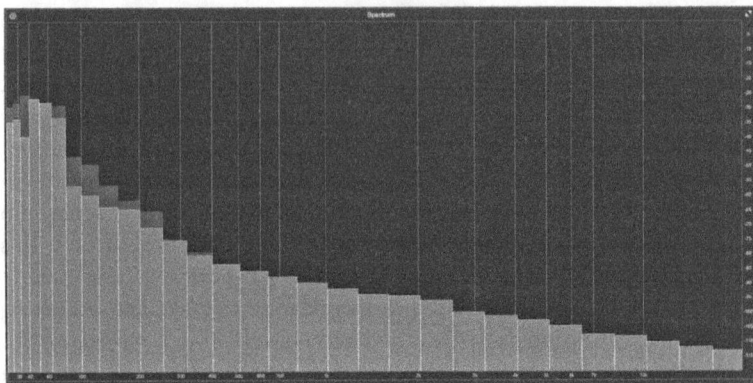

Figure 16.1 Spectrum analysis of Gaspar Noé's *Irréversible* (2002) using the audio analysis software iZotope Insight 2. The data was captured during the opening credits of the film at the timecode 00:01:04:15 (hour:min:sec:frame), using one-third-octave band frequency filters. One-third-octave band measurement is a standard for scientific instruments when determining Noise Criterion and Noise Rating Curve. The horizontal axis represents frequency from 20 Hz to 20 kHz left to right, and the vertical axis represents decibels relative to full scale from -120 dBFS to 0 dBFS bottom to top. This screenshot was made on 23 February 2021.

"spiked" with a constant 27 Hz tone used by the police in crowd control. The sound induces feelings of nausea and vomiting and was inserted into the film by Noé for precisely these reasons'.[25] However, using spectral analysis software, it is easy to illustrate there is no single stable fundamental frequency. Furthermore, spectrogram analysis of the soundtrack clarifies that the sub-bass does not play throughout the film, its duration is not ubiquitous, nor is it at a constant amplitude for thirty minutes or sixty minutes, as claimed. Nevertheless, the affective force of the sub-bass is in no way diminished. During the narrative block set inside the sex club called the Rectum, the sub-bass drone becomes musical with an ostinato pattern using glissando to create a doppler effect. The sound here is unstable and unsettling, designed with the intention of causing bodily affections. The sub-bass modulations, along with the harmonics and overtones, coalesce with other sound elements to create a maelstrom of sonic affect. Neil Lerner explains that specific sounds in cinema can operate as 'assaultive blasts' that 'affect us at a primal level, perhaps instinctually taking us back to a much earlier time when the ability to perceive a variety of sounds alerted us (as a species) to approaching predators or other threats'.[26] Sub-bass operates as such an alerting, assaultive blast and the simplistic assertion is that the lower the tone, the bigger the danger. It is here the film plays on a logic of sensation between its low tone modulations and the audience, through an omnipresent, enveloping force of vibration. The sub-bass is acting on the central nervous system of its audience where sound waves of significant magnitude modulate bodily organs in sympathetic resonance with the narrative violence. Throughout the film the sub-bass content changes, with intermitting pauses and pitch shifts. It is most prominent during the tiles, the opening segment outside the Rectum and the narrative block located inside the Rectum, which spans over twenty-five minutes. The consequence of this is a sonic-affective bombardment before the film's most harrowing scene yet to come, the brutal rape of Alex. Thereafter the low-end noise is an intermittent feature. Nevertheless, there is significant energy in the sub-bass region for a large portion of the film. It is present during all transitions and plays constantly throughout the rape segment (see Plate 13). Curiously, during the rape it is pitched higher with the fundamental frequencies ranging from 46 Hz to 59 Hz (see Figure 16.2).

As can be seen in the figures, the sub-bass content in *Irréversible* is significant. Compared to a typical Hollywood film of the same era using the same type of analysis software there is considerable difference in the sub-bass weighting.[27] For example, during the opening minutes of the film, the sub-bass content is 70 dBFS (decibels with reference to full scale) louder than the eighth octave band (1420 Hz–2840 Hz) (see Figure 16.3) where dialogue typically has presence. While this analysis illustrates the frequency content on the DVD with reference to dBFS, it tells us nothing about the audible acoustic playback of that content – it was performed digitally and absent of acoustic energy in space. In understanding the affective quality of the sub-bass in this film, and sub-bass as an affect more widely, it is critical to acknowledge the degrees of acoustic intensity in which it is experienced. For it is acoustic intensity that gives sub-bass its physicality earlier noted by Morgan. In a fundamental way, sub-bass as an affect is based on its amplitude of vibration and potential for creating affective resonance. Its affection is determined by its power and intensity. With respect to *Irréversible*, each listening event will have a different sonic quality due to the playback system and the acoustic environment composed of interacting, absorbing and resonating bodies. While this can be said of all sound, sub-bass is more affective and affected by intensity. Where and how one hears the film shapes its ability for low-frequency propagation and corporeal affection. Listening through laptop speakers is a rather different experience to hearing it through a cinema sound system capable of playing back the full spectrum of audible sub-bass. This is to say, the listening

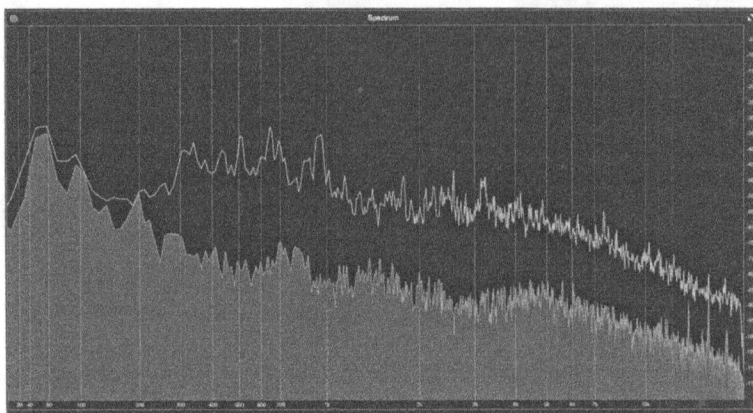

Figure 16.2 Linear spectrum analysis of Gaspar Noé's *Irréversible* (2002) during the rape scene. The data was captured with iZotope Insight 2 at timecode 00:49:01:12, using the constant bandwidth FFT (fast Fourier transform) setting. The brighter white line indicates the maximum peaks over a five-second period. The horizontal axis represents frequency from 20 Hz to 20 kHz left to right, and the vertical axis represents decibels relative to full scale from -120 dBFS to 0 dBFS bottom to top. During this section, the sub-bass is pitched higher than in other parts of the film, with the fundamental frequencies oscillating between 46 Hz and 59 Hz. This screenshot was made on 23 February 2021.

context and sound pressure level experienced are instrumental in sub-bass's affective quality. This is crucially what Cox's sonic materiality encapsulates: sound is a relation, in Cox's words, it is the possibility of change within the material relations between bodies. Due to its physicality, sub-bass can have a physiological effect on the human body – the affective force of which is in an exponential relationship with amplitude.[28] Therefore, the material formation of low frequency as an affect is critically understood by its power as a degree of intensity, and it is here that sub-bass can be thought through the Deleuzian framework offered.

To recap, for Deleuze, affection is a state and affect is a transition from one state to another governed by a degree of intensity. Bruce Baugh explains that within Deleuzian thought: 'Power is physical energy, a degree of intensity, so that every increase or decrease in power is an increase or decrease in intensity . . . Every transition from a greater to a lesser intensity, or from a lesser to a greater, involves and envelops the zero intensity with respect to which it experiences its power as increasing or decreasing.'[29] While Baugh is

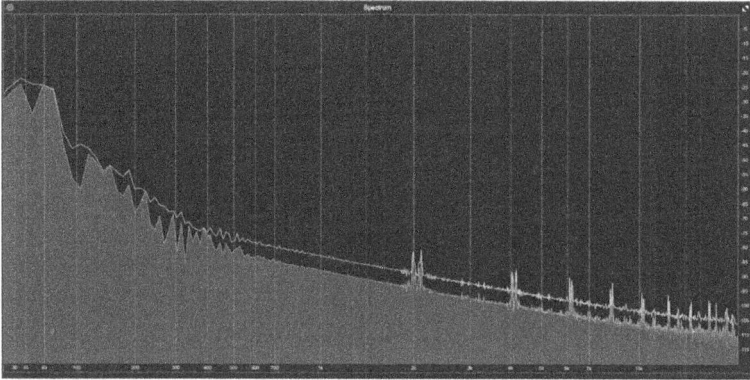

Figure 16.3 Linear spectrum analysis of Gaspar Noé's *Irréversible* (2002) during the opening credits. The data was captured at timecode 00:01:47:07, using iZotope Insight 2, set to measure with constant bandwidth FFT (fast Fourier transform) analysis. The horizontal axis represents frequency from 20 Hz to 20 kHz left to right, and the vertical axis represents decibels relative to full scale from -120 dBFS to 0 dBFS bottom to top. During this segment, the sub-bass content is 70 dBFS louder than the eighth octave band (1420 Hz–2840 Hz) where dialogue has the most presence. This screenshot was made on 23 February 2021.

explicitly referencing the concept of death in Deleuze, the same can be said of sub-bass and the degrees of power determining the intensity of affection in a receiving body. It is in the degrees of power where sub-bass as affect can be uncloaked. What makes sub-bass unique is the register it operates in; this is the domain where sound as affect unfolds in logarithmic magnitudes of bodily experience. It is its shear physicality upon other bodies. When sub-bass breaks thresholds of sensation it does so at degrees of intensity many times more than other registers. Audible sub-bass has a magnitude and amplitude that dwarfs mid-range and high frequencies. We can say that all sound has the potential for affection; at a fundamental level, sound is the transmission of energy causing variations between bodies. However, sub-bass is more capable of causing bodily resonance due to its physicality and intensity.

When auditioned at amplitudes that cause acoustic resonance, sub-bass affects all types of bodies, easily moving through corporeal bodies, but also through the fabric and foundations of the buildings that contain them. It effortlessly creates material relations. As the work of sound artist Mark Bain illustrates, low frequency can cause buildings to vibrate and create 'transient architecture'.[30] It is this potential of intensity, the power to move buildings, that marks sub-bass as a specific sound affect. When acting on human bodies it can easily modulate organs. Sub-bass as an affective force is an oscillating intensity of becoming, whereby there is a passage from one configuration to another, causing the transmission of affect and creating sympathetic vibrational movement. In this becoming – a becoming of analogous vibrations – bodies as receivers become participants and further shape the sub-bass within the sonic arena. Consequently, these absorbent and reflective bodies also become affective bodies, acting on and reshaping the sub-bass vibrations. Compared to other frequencies, sub-bass is more pronounced in this relation of sonic vibrational affect because of the spaces that try to contain it. When contained within rooms of any type, low frequencies are prone to creating standing waves, thus causing a further affective chain between architectural bodies, inanimate bodies and corporeal bodies in a relation of vibratory resonance. Furthermore, understanding sound and sub-bass as a material relation has nothing to do with representation, and everything to do with sensation. Sub-bass is a gradient of intensity involved in affection through the modulations of affected bodies. It should be understood as a dynamic flow, with varying intensities that are tied to sensation, registered as an intensity rather than something bound in signification. Consequently, sub-bass has no specific semiotic meaning, nor has it a prescribed affective outcome, it possesses the ability to generate both good and bad vibes. Like any sound it can cause affect, but because it is the lowest audible register of human hearing, it operates in the domain of an exponential intensity of vibration. As noise, sub-bass vibration is unwanted and therefore troublesome, as music it presents fullness and enclosure to propel bodies to move in compliant resonance. In each case it is its intensity that gives sub-bass its physicality and affective force.

In considering how sub-bass transmits affect *Irréversible* is a particularly poignant example because of its overt use of the sonorous-affective force. From the spectral analysis performed, we can deduce that there is an adulterated intentionality behind the film's sound design through its employment of sub-bass frequencies explicitly invested in affect. As shown, graphical representations of the soundtrack reveal significant and extended low tones. However, hearing the film in ideal conditions is critical in its production of affections. The same is applicable to all audition of sub-bass. The Cannes Film Festival premiere screening represents an ideal listening experience, but also an event that has shaped the film's meaning thereafter due to the notoriety it gained. It is fair to say that the extended power of the sub-bass in the film is deliberately designed to provoke its audiences' mechanoreceptors, using vibratory pressure to create physiological somatosensory effects – this is sound as affect, body as affected. The cultural reception of the film speaks to how affects might operate as sound in a loose sense – what sounds did the audience make in response to the low-frequency oscillations? If our understanding of sound is a material flux composed by vibrational force, the affections in the body caused by low-frequency vibration just might have a corporeal resonance.

Notes

1 This can be illustrated using Robinson-Dadson's Equal-Loudness Contour, an ISO standard (International Organization for Standardization) created to measure frequency's relationship to perceived loudness. For a full explanation of physics involved in loudness perception measurement, see Alton F. Everest, *The Master Handbook of Acoustics*, 4th edn (New York: McGraw-Hill, 2001), 41–81.

2 A study in 2019 found an upward trend of extended bass frequencies in increased amplitude in popular chart music from 1955 to 2016. The authors conclude that this could be due to changes in technology and style 'but based on links between bass and movement, it is likely a widespread technique to increase engagement and contribute to chart success'. Michael J. Hove, Peter Vuust and Jan Stupacher, 'Increased Levels of Bass in Popular Music Recordings 1955–2016 and Their Relation to Loudness', *Journal of the Acoustical Society of America* 145, no. 4 (2019): 2247–53.

3 Steve Goodman, *Sonic Warfare: Sound, Affect, and the Ecology of Fear* (Cambridge, MA and London: MIT Press, 2012), 79.

4 Paul C. Jasen, *Low End Theory: Bass, Bodies and the Materiality of Sonic Experience* (New York; London; Oxford; New Delhi; Sydney: Bloomsbury Academic, 2016), 9.

5 Geoff Leventhall, 'Low Frequency Noise. What We Know, What We Do Not Know, and What We Would like to Know', *Journal of Low Frequency Noise, Vibration and Active Control* 28, no. 2 (2009): 79–104; M. Schust, 'Effects of Low Frequency Noise up to 100 Hz', *Noise and Health* 6, no. 23 (2004): 73; Yukio Takahashi et al., 'A Pilot Study on the Human Body Vibration Induced by Low Frequency Noise', *Industrial Health* 37, no. 1 (1999): 28–35; Yukio Takahashi, 'A Study on the Contribution of Body Vibrations to the Vibratory Sensation Induced by High-Level, Complex Low-Frequency Noise', *Noise and Health* 13, no. 50 (2011): 2.

6 Frances Morgan, 'Darkness Audible: Sub-Bass, Tape Decay and Lynchian Noise', in *The End: An Electric Sheep Anthology*, ed. Virginie Sélavy (London: Strange Attractor Press, 2011), 200.

7 Alexandra Heller-Nicholas, *Rape-Revenge Films: A Critical Study* (Jefferson, North Carolina; London: McFarland, 2011), 21.

8 Sergei Eisenstein, *S. M. Eisenstein: Selected Works*, Vol. 1 Writings, 1922–34, ed. and trans. Richard Taylor (London: BFI Publishing, 1988), 3.
9 Matt Bailey, 'Noé, Gaspar', *Senses of Cinema* (2003), https://www.sensesof-cinema.com/2003/great-directors/noe/ (accessed 9 November 2020).
10 Tom Gunning, 'The Cinema of Attractions: Early Film, Its Spectator, and the Avant-Garde', in *The Cinema of Attractions Reloaded*, ed. Wanda Strauven (Amsterdam: Amsterdam University Press, 2006), 382.
11 Tim Palmer, 'Style and Sensation in the Contemporary French Cinema of the Body', *Journal of Film and Video* 58, no. 3 (Fall 2006): 22–32; Tim Palmer, 'Under Your Skin: Marina de Van and the Contemporary French *cinéma du corps*', *Studies in French Cinema* 6, no. 3 (2006): 171–81.
12 *BBC News*, 'Cannes Film Sickens Audience', 26 May 2002, http://news.bbc.co.uk/1/hi/entertainment/2008796.stm (accessed 1 November 2021).
13 Christoph Cox, 'Beyond Representation and Signification: Toward a Sonic Materialism', *Journal of Visual Culture* 10, no. 2 (2011): 156.
14 Cox, 'Beyond Representation and Signification', 156.
15 Cox, 'Beyond Representation and Signification', 156.
16 Cox draws on multiple texts from Nietzsche's oeuvre to trace the development of becoming in Nietzsche's thought.
17 Cox, 'Beyond Representation and Signification', 157.
18 Cox, 'Beyond Representation and Signification', 157. It is worth mentioning that while Cox never uses the term affect, it is further implied in his reading of Nietzsche's 'will to power' as a transformative process composed of causality and effectivity.
19 Gilles Deleuze and Felix Guattari, *A Thousand Plateaus: Capitalism and Schizophrenia*, trans. Brian Massumi (1980; Minneapolis and London: University of Minnesota Press, 1987), 256.
20 Deleuze and Guattari, *A Thousand Plateaus*, 305.
21 Eugene B. Young, Gary Genosko, and Janell Watson, *The Deleuze and Guattari Dictionary* (London: Bloomsbury Academic, 2013), 166.
22 Gilles Deleuze, *The Fold: Leibniz and the Baroque*, trans. Tom Conley (1988; London: The Athlone Press, 1993), 77.
23 Gilles Deleuze, *The Logic of Sense*, trans. Mark Lester (1969; London: The Athlone Press, 1990), 8.
24 Bailey, 'Noé, Gaspar'.
25 David Davies, 'Watching the Unwatchable: *Irréversible*, Empire, and the Paradox of Intentionally Inaccessible Art', in *Suffering Art Gladly: The Paradox of Negative Emotion in Art*, ed. Jerrold Levinson (Basingstoke and New York: Palgrave Macmillan, 2014), 252.
26 Neil Lerner, ed., *Music in the Horror Film: Listening to Fear* (New York and London: Routledge, 2010), ix.
27 While many films from this era certainly feature high energy sub-bass frequencies at times of high intensity or spectacular moments, what sets *Irréversible* apart is the duration for which low-frequency noise is present.

28 Medical studies on the effects of low-frequency noise confirm this. See the earlier footnote on the effects of low-frequency noise documented by researchers Leventhall, Schust and Takahashi.

29 Bruce Baugh, 'Death', in *The Deleuze Dictionary*, ed. Adrian Parr, rev. edn (Edinburgh: Edinburgh University Press, 2010), 64.

30 Mark Bain, 'Transient Architecture: On the Making of the Live Room', *Thresholds* 16, no. Spring (1998): 50–3.

Squish, squelch, shlshlshlurpppp

Norie Neumark

Introduction

Squish, squelch, shlshlshlurpppp[1] . . . the nearly inaudible sound of wormy compost . . . its more-than-human voice. As rotting compost and wriggling worms speak . . . *squish, squelch, shlshlshlurpppp* . . . they arouse human bodily affects. These human bodily affects, in turn, sound out, voicing themselves. Some people recoil with abject disgust, *gasping* their deathly horror; others are joyfully lured closer, *humming* their connection; while still others simply turn away in numbed indifference, *hrrrumphing* their diffidence. Through new materialism, critical animal studies and sound theory, I will listen to wormy compost's more-than-human *voice* – a sound calling out for a response – as it evokes human affects.[2] I will ask, how do the more-than-human voices of rotting compost and wriggling worms stirs affects in their human listeners? How does perceiving the affective relations between these human and more-than-human voices complexify our understandings of the circulating loop between sound-producing affects and affects operating as sound?

As I remember and respond to the affective lure of wormy compost's voice, I'm provoked, too, to reconsider broader aesthetic and political understandings of human/more-than-human relations. Making these aesthetics and politics audible is the work of art and its practice; and to explore this, I will consider my own art practice, inspired along the way by writers and theorists such as Vinciane Despret, Eva Meijer, Jane Bennett and Kathleen Stewart. For all of these writers, the connection of affect and ethics and sound resonates, sometimes in the foreground, sometimes in the nearly inaudible background – like wormy compost itself. *Squish, squelch, shlshlshlurpppp.*

Inclining in . . .

Let me first state my inclinations. I am a compost devotee and a worm lover. I incline to listen joyfully to their nearly inaudible voices – voices that move me affectively as they lure me in. As a sound artist, I'm inclined to listen and to respond through my work. This is more than a metaphorical 'listening', politically and ethically important though that is.[3] This is a literal, bodily, affective and sonic inclination.

An urge to make literal what Meijer would point to as 'listening' to worms was one of the concerns grounding the *worms* project I developed with artist/ collaborator Maria Miranda. Alongside the worms themselves, we were especially inspired by Despret's vital understanding of human-non-human animal relations as a mutual *attunement* – a passionate, bodily *with-ness*. For Despret, such relations of attunement depend on the availability of the bodies to each other, to practices that transform human and non-human animal bodies to the affects that move them.[4]

To burrow into this attunement, we wanted to 'work with' worms, to 'follow' them.[5] We decided on a durational artwork – as we fed the worms, they transformed 'dead' matter into live soil, into compost. We began the project soon after we had moved from an apartment to a house, where for the first time we had a backyard, vegetable patches and a worm café – a very Naarm/ Melbourne inner suburban way of life. We hadn't thought about worms before, but we were quickly enchanted. The more we fed them, the more fascinated we became. We discovered shared passions, like the love of coffee, which animated us all. (*Ahhh*, surge of a coffee high, for those of you who share this, *ohhh*, shudder of disappointment for those who don't). We were further drawn to work with the worms when we sensed an affinity between our commitment to recycling and their composting/transformational skills. We were moved by a strange but intimate emotional connection to the earthworms – little critters who are so vital to everyday life and to the existence of humans, non-humans and the earth. They are normally overlooked in the background, even despised, but, for us, they evoked joy . . . *YES*.[6]

As for many artists, a crucial question presented itself before we began the project – how to work with non-human animals ethically and politically as well as aesthetically – in ways that don't exploit their vulnerability? How, in other words, to 'follow' them, rather than using them or putting them on display? As artists, we decided to try to attune to the worms through listening to their voices, feeling their moving voices provoke our own affective voices. Through our headphones attached to a microphone in the worm café, we listened intently. We let our bodies and voices *amplify* their voices. *Shlshlshlurpppp*, they said . . . *Shlshlshlurpppp*, we responded.[7]

Encountering the worms during the project, I was moved to read, to think and to write.[8] I noticed even Charles Darwin recognized worms' intelligence, creativity and sociability: 'They perhaps have a trace of social feeling, for they are not disturbed by crawling over each other's bodies, and they sometimes lie in contact.'[9] Before I actually attuned to worms, I would have cringed at this image – crawling, slimy, creepy – *errrkh, eesehhhshh*. But now I feel differently – affectively and sonically. Now when I lean in to listen, I apprehend an alluring voice co-composed of the *whirring* coffee machine that grinds the morning coffee we share, the *crunching* toast we're all *munching*, the *shlurping* food as it goes into the compost – together voicing an assemblage that calls me to listen to what an art collaboration with worms might be and might say. *Ahhhh, Yes.*

Inclining in . . . In making the work, I experienced my inclination literally, physically – as an affective force produced by sound and operating as voice that pulled me closer to the worms. I turned to political philosopher Jacques Rancière to think about the aesthetic/political implications of this force, which moves bodies affectively. Rancière proposes that there is a 'commonsense' regime of the senses or sensorium upon which the givenness and habits of our society rest: 'the distribution of the sensible, or the system of divisions and boundaries that define, among other things, what is visible and audible within a particular aesthetico-political regime'.[10] This is what, for him, makes art inherently political – since the social, cultural and political *status quo* operates through that invisibility and inaudibility.[11] Through its affective force, then, art can make us notice things that we have come to take for granted, things we have become accustomed to not noticing and, I would add, to deny.[12] Art can disturb our habits, throw daily life and its groundings into relief and thus open up new ways to think about what we do. Things we feel, affectively, but don't register consciously or intellectually – these things are the ground that art can work on ethically and politically, to motivate recognition, response and/or refusal of denial – for the artists as well as audiences.

As artists, Maria and I participated in a political and ethical foregrounding of critters and affective relationships – relationships which are usually unnoticed, though not unfelt nor unspoken. *Shlshlshlurpppp*.[13] As wormy compost's voices moved us affectively, forcefully, we were provoked to practice care and response-ability. We hoped, too, that these small voices – theirs and ours – might stir others as well. Perhaps they might motivate an ethical response to pervasive anthropocentric exploitation of vulnerable non-human animals – affected by and expressing our fundamental entanglements.

Hmmm, entanglements

Hmmm, entanglements. So much that is written about affective relationships with non-human animals relies on the figure of entanglements. But this is not an untroubled figure. It calls for untangling. As does that other (over-used?) figure, 'becoming-animal', to which I'll turn first. Becoming-animal is one of Gilles Deleuze and Félix Guattari's (henceforth D&G) many *becomings*. Like so much of D&G's work, the concept seems purposely slippery, or can I say wriggly? It seems to vibrate somewhere between or beside the literal and metaphoric, since D&G are trying to avoid ontological states of being. They talk philosophically about becoming-animal as movements by *contagion*, as a way of thinking movements that are not about the more familiar relations of pity, identification, analogy, imitation, representation, resemblance, or reproduction. In this Spinozan vein, they are invoking forces and 'a proximity "that makes it impossible to say where the boundary between the human and the animal lies"'.[14] This concept resonates more for me when digested by philosopher Christoph Cox, who takes it into artful and sonic zones. Cox reckons becoming-animal is the practice of being 'drawn into a zone of action or passion that one can have in common with an animal. It is a matter of unlearning physical and emotional habits and learning to take on new ones.'[15]

From Cox, I sense becoming-animal as entering (should I say worming into?) a shared affective and productive zone, to experience common capacities with non-human animals – an affective zone animated by sound affects, I would add. And in this movement, Cox explains, we can experience new physicalities, new emotions, and new relations with others and with the world. By engaging with particular artworks, Cox sidesteps what seems like a tendency for abstraction in D&G's work. Perhaps in this way Cox helps rescue becoming-animal from what feminist scholar Donna Haraway criticizes as D&G's 'disdain for the daily, the ordinary, the affectional rather than the sublime'.[16] And for me, the worms always/already provoked just such affect and affection – after all, what could be more 'daily' and 'ordinary' than worms – with whom we are connected by sound affects and affection? *Squish, squelch, shlshlshlurpppp*. The 'ordinary' is certainly not to be ignored, as the worms, like anthropologist Kathleen Stewart, insist. In her writing on ordinary affects, Stewart speculates on the 'immanent force' of 'the complex and uncertain objects that fascinate because they literally hit us or exert a pull on us . . . Ordinary affect is a surging, a rubbing, a connection of some kind that has an impact.'[17] *Ahhh, yes*, Kathleen, I am literally pulled, sonically, affectively, to follow these ordinary critters. Entangled by sound affects?

Hmmm, entanglements . . . and their untanglings. Like 'becoming with', 'entanglement' is a powerful and appealing figure for many in animal studies; however, it, too, risks occluding many human/non-human animal relations that we don't hear (or deny) while we roll this poetic figure round our tongues.[18] Paula Arcari, Fiona Probyn-Rapsey and Hayley Singer, in particular, have critiqued entanglement as a celebratory, romanticizing, inadequately politicized figure that has distracted academics across a range of disciplines from critically noticing the commonplace instrumentalizing and commodification of non-human animals. As they note: 'in contrast to the ethos of multispecies "entanglement" and "becoming with" that typically animates this research, large numbers of animals "entangled" in the machinations of our cities constitute a "nature" that remains mostly unseen.'[19]

Reading their critique, I realize that while the figure of entanglement might direct us to listen to how a lamb in a petting zoo *bleats* joy for the child who leans in to pet it . . . *ahhh* . . . the concept doesn't necessarily attune us to the pitiful, but unpitied, *cries* of the lamb in the abattoir, cries that are not to be heard or felt by that child . . . *ohhh.*

Nonetheless, as I watch worms intertwining with each other, and as I respond affectively and sonically, I do feel a pull which feels like entanglement – inescapably entwined and connected, at the same time to these others. And I wonder if the figure of entanglement might still have useful conceptual force, making us notice an intense otherness that calls out to us affectively for response. I wonder if worms and their composty entanglements might, particularly, open our senses and bodies to other, unfamiliar relations, as Meijer seems to suggest with her 'worm politics'. As she proffers, it is 'problematic to only focus on those animals who are most like us. Because earthworms are so different to humans, and because we know little about them, our relations with them do raise many questions'.[20] And one of these questions is about affective entanglements and knowledge – the knowledge/power relations that help shape and are shaped by these affects, these sound affects. *Squish, squelch, shlshlshlurpppp.*

As I think about Foucauldian knowledge/power and affect, I'm painfully reminded of a disturbing event that haunts me still. I retell this story inspired by Kathleen Stewart's speculative writing that recognizes 'how moving forces are immanent in scenes, subjects and encounters'.[21] The event occurred during the launch in my local park of a Kathy Kolowko's statue of a giant, golden earthworm, titled *Unsung Hero* (20 October 2019). The launch included a 'Worm Party', at which the Gould League promised a free worm composting session and information stall. I recall going over to the stall and recoiling as I witnessed children being encouraged to poke and prod worms in the name of 'science' and 'education'. I felt my chest tighten, my body hot with anger, itching to explode in a rant about the cruelty. *NO.* I tried to reason, *gulp*, and alert the stallholder to the alienated cruelty, to which the children were being incited, but I couldn't keep my emotion from my voice. 'That's how children learn', she asserted testily. I could barely restrain from poking and prodding *her* myself, retorting, 'I'm just trying to learn about science.' *Hrrumpf. Grrrrr.* In respect for the artist, I refrained from creating a scene, even though I was deeply dejected by an educator's prodding children to poke those vulnerable worms: this was, after all, the launch of a statue in honour of earthworms, a statue that put an earthworm on a pedestal to be admired – not to be touched and, certainly, not to be assaulted in the name of 'learning'. It was meant to honour the worm and teach about ethical, caring composting.[22] But I do now wish I had followed my inclination to liberate those suffering worms. *Mmmmhh.*

Remembering this incident, I feel the affect yet again – the anger shaking my body as it was assaulted by the *moral tone* agitating the 'educator's' voice. As I hear, again, the *No* escaping my lips, I am impelled to think, to wonder, how can I understand this instrumentalizing of animals as objects for 'education'? I tremble with fury, *urrgghh,* just as when I hear people defending zoos[23] as scientific and educational (what those non-human animals in zoos – ripped from their own place and kin, for 'our' educational benefit – learn about human animals is the real question). I shudder at the high moral ground of science amplifying that stallholder's voice. And once more I find myself voicing *No* to the human exceptionalism, which would assert its right to extract knowledge from non-human animals' vulnerable bodies.

This affective cringe, this shudder, and the *ohhh, no* they produce lure me, now, to lean back and reflect on ethics. Theorist of environmental, multispecies ethics, Mick Smith proposes: 'There is a sense in which the Other always calls us out of the world with which we are familiar to insist that there is more, infinitely more, than initially appears to be the case, more than we know, see, or understand.'[24] In place of the hubristic demand to know, to fully grasp or comprehend the individual self of an Other, he prefers an ethics of 'fellow-feeling'.[25] For Smith, this ethics rests 'in appreciating the not entirely comprehensible ways in which . . . individuals . . . constitute a part of a community of myriad beings which appear to each other in all kinds of ways, as commensual, as mutualistic, as parasite, as prey, as resources, as co-evolved and evolving beings'.[26] I lean back in, *sigh, ahhh, yes* – listening to the affective voice of more-than-human others induce just such an ethics.

Fellow-feeling . . . affective relationships

I am affectively moved by fellow-feeling with the worms – I sense it as and through an affective relationship, *ahhh.* This affective relationship is intimate. This intimacy is not that of lovers or family or even friends. It's more a neighbourly intimacy – close, contingent, and/but so often troubled – *yes* and/

but *no*. Intimacy, indeed, is another tricky one. So let me explore intimacy further, when it is entangled with affect.

Feminist technoscience scholar Tara Mehrabi (who herself spent a year intimately relating with fruit flies in a science lab) explores 'multispecies intimacy'. She works with Helga Sadowski's understanding of an 'affective intimacy' between human and non-human animal bodies: 'Through affective intimacy bodies perform one another, animate each other simultaneously . . . therefore, affective intimacy is an embodied performative relationality that "is not necessarily related to feelings and emotion, although they might emerge in the mix of affects, which is affection".[27] Following animal studies scholar Kathryn Gillespie, Mehrabi perceives multispecies intimacy as 'entangled with empathy . . . a blend of emotion and cognition in which we enter into relations with others, animals, the environment and the world. We are therefore *called upon* to be responsible and to attend to other's diverse modes of *vulnerability* and *situated needs*.'[28] This, I'm thinking, was the intimate affective call of the worms at that so-called party – *uunngg*, I shiver, even again now, that I did not respond effectively to their affective call.

And then, there's another face of intimacy, when we're entangled in encounters that, say, stir disgust. Again, Mehrabi considers the affective relations of intimacy and ethics:

> I also ask what kinds of empirical material one can collect by thinking through the concept of intimacy as affective encounters that are not inspired by the warm feelings of love and companionship that fluffy animals such as dogs trigger . . . but instead are saturated with the repellent intensity of disgust and discomfort of being intimate with the abject, here fruit flies. Finally I ask whether, in thinking through *multi-species intimacy*, one can rethink ethics as a relational process of becoming with that extends beyond the comfort zone of 'nearby geographies'.[29]

Urhhghhh . . . yukkkkhkh Let's stay with the trouble of disgust a bit longer, maybe a lot longer, since we are talking wormy compost after all. *Urhhghhh . . .* disgust . . . an 'intensification of movement', from which we 'recoil', says feminist theorist Sara Ahmed. For Ahmed, disgust is a productive affective (or in her terms, emotional) relationship – intimate, proximate, sensuous.[30] She responds to it as a form of vomiting . . . to abjectly expel something already digested and incorporated in the body.[31] Ahmed makes me wonder, is the disgust some people feel around worms to do with the way these critters ingest, digest, eject our food . . . is that part of the abject creepiness animating that recoil? Retching . . . *urhhghh*

This brings me, finally, to my affective affection for wormy compost: my joy in decomposition, rotting, digestion, and how they, as Mehrabi might put it, are 'queering the very understanding of intimacy'.[32]

Squish, squelch . . . It's still life

Still life is a forceful figure of 'intensity' that pulses for Kathleen Stewart: 'A still life is a static state filled with vibratory motion, or resonance. A quivering in the stability of a category or a trajectory, it gives the ordinary the charge of an unfolding.'[33] Still life is what Maria Miranda and I sense when we encounter our wormy compost. *Squish, squelch* . . . the nearly inaudible sounds of the ordinary but extraordinary processes of compost's decomposition, rotting, digestion. And the joyful sounds of those who respond to their affective lure . . . *ahhh, ohhh, yes.* These are familiar to Maria and me and they led us to a recent sound artwork around compost, which we called *It's Still Life.* For that project, we recorded the responses to compost by neighbouring community gardeners. We asked them to gently dig into the compost, to turn it, to listen to its *squish, squelch* voice and to describe their affective response as they did so.[34]

What we sensed in their voices, attuning to the *still life* that is compost, was an intense, intimate affective encounter with liminality, life and death, life in death. Compost's nearly inaudible voice disturbed some of those gardeners with the palpable, quivering horror of death and decay. Yet, intensified with the nearly inaudible voice of the commingling worms, who find life in compost and give it life, the compost moved others in tune with the wonder of the critters' own sociality and creativity and transformative capacities. 'It's alchemical', said one of the gardeners in delight and awe. So let me, one final time, follow the turbulent thinking that these affective voices provoke, following alchemical doing and thinking into its dirty, noisy, bodily depths as I tune in to the voices of alchemists.

Alchemy. Years ago I made a radiophonic work about alchemy – urged by the way it speaks to so many artists as a way into their process.[35] I remember still my startle and pleasure as I listened to anthropologist Jonathan Marshall and to other contemporary alchemists speak of alchemy's slippery transformations (does this remind you of worms?). *Ohh . . . Ahhh.* Alchemy, Marshall told me, involves instabilities which begin a transformative process. It is not about fixity, regularity and reduplicability but about things which are temporary or may only happen once in someone's lifetime. It's about particularity not generalized abstraction.[36]

Alchemical doing/thinking disrupts the seemingly unquestionable (religious?) primacy of modern science. It resonates with essayist Jonathan Crary's discussion of how scientific discourse, in the nineteenth century, contributed to a reconfiguration, a separation of the senses with the disassociation of sight from touch and the increasing abstraction of sight.[37] Crary recognized that moment of the remaking of vision and its objects as important to the remaking and abstracting of the modern bourgeois individual – 'remaking the individual as observer into something calculable and regularizable and of human vision into something measurable and thus exchangeable'.[38] The inexorable logic of *capitalist* modernization, which Crary describes as driven by a logic of the same, the making exchangeable and circulatable that which is singular, is the basis of science's extractivist instrumentalizing. And this is the modern science, which presumes that animals 'belong' to them as their 'proper subjects'.[39] This is the science, with its 'knowledge-production practices'[40] that justified the 'education' into human exceptionalism that arrogantly disturbed the singular, specific, vulnerable beings at the worm 'party'. (*Ohhh*, I quiver still with grief as I remember the sound affects at that event.)

With a new materialist approach not limited by such scientific thinking, theorist Jane Bennett, too, finds interest in alchemy and its affects. Writing of the alchemist Paracelsus, she discerned: 'For him, enchantment is not only a property of the natural world – it is also the joyful human mood that results from a special way of engaging that world. Enchantment as a mood requires a cultivated form of perception, a discerning and meticulous attentiveness to the singular specificity of things.'[41] Like Jonathan Marshall, Bennett noted that listening, or eavesdropping, is crucial to alchemical practice – which returns me to the question of how sound affects provoke ethics and politics.[42]

Bennett's writing on enchantment opens sonically onto this question. Her understanding of generous behaviour to others (human and more-than-human) rests on a recognition of the need for a motivating energy for ethics – an affective energy. Such an energy can come from enchantment. Bennett particularly notes the 'chant' animating enchantment – the sound that provokes and evokes it:[43] '*Enchantment* is a peculiar kind of mood, often induced by sound (the *chant* in enchantment). To be enchanted is to be both charmed and disturbed: charmed by a fascinating repetition of sounds or images, disturbed to find that, although your sense-perception has become intensified, your background sense of order has flown out the door.'[44] This is the affective enchantment whose intense and moving sonic forces circulate as sound affect, motivating political and ethical responses. *Yes/No.*

Squish, squelch, shlshlshlurpppp

I have listened to these enchanting and disturbing voices of sound affects, sound circulating as affect and affect circulating as sound, as I followed wormy compost. I asked at the beginning, how these more-than-human voices of rotting compost and wriggling worms stirred affects in its listeners. As I have pursued this listening practice, as I have been moved by these voices, I have come in the end to attune to the still life, the life and death, affectively animating these voices. What I have been finally drawn in to listen to with wormy compost is that *it's still life*. Listening to the *squishy*, *squelchy* voice and sensing the sound affects moving into and out of my own body, I have been responding to the ethical and political importance of attuning to the affective voices of the most ordinary critters.

Notes

1 I will use italics throughout for the sound and voice of affects.
2 Norie Neumark, *Voicetracks: Attuning to Voice in Media and the Arts* (Cambridge, MA: MIT Press, 2017).
3 Eva Meijer, *When Animals Speak: Toward an Interspecies Democracy* (New York: New York University Press, 2019), 161–3.
4 Vinciane Despret, 'The Body We Care For: Figures of Anthropo-zoo-genesis', *Body and Society* 10, nos 2–3 (2004): 125–70.
5 Previously we wrote about this as 'collaboration' but now, in recognition of the fact that we never sought nor received the worms' consent, we're using 'work with' and 'follow', following Jacques Derrida and Katve-Kaisa Kontturi. The title and text of Derrida's book plays with the French for *je suis*, which means *both* 'I am' and 'I follow'. See Jacques Derrida, *The Animal That Therefore I Am*, trans. D. Wills (2002; New York: Fordham University Press, 2008), passim.
6 Following Howard Morphy's discussion of art in the Arnhem Land region of North Australia, anthropologist Deborah Bird Rose discusses the Yolngu term, *bir'yun*: '*Bir'yun* is the shimmer, the brilliance, and the artists say, it is a kind of motion. Brilliance actually grabs you. Brilliance allows you, or brings you, in the experience of being part of a vibrant and vibrating world.' This lure, this capture brings joy and knowledge. See Deborah Bird Rose, 'Shimmer: When Everything You Love Is Being Trashed', in *Arts of Living on a Damaged Planet: Ghosts and Monsters of the Anthropocene*, ed. A. L. Tsing, H. A. Swanson, E. Gan, and N. Bubandt (Minneapolis: University of Minnesota Press, 2017), G53–54. 'Trees call out to flying foxes in languages of color and scent, and flying foxes respond with gusto.' '*Yes*,'

she says, they say (Rose, 'Shimmer', G60). Rose's affective, shimmering *yes* chimes, I sense, with artist Yoko Ono's 1966 work, *yes* (Rose, 'Shimmer', G60–61).

7 Norie Neumark and Maria Miranda, *Waiting* [online video component of art installation], MoreArt, Melbourne, Australia, 2017, https://vimeo.com /247735081 (accessed 28 January 2022).

8 Norie Neumark, *Working with Worms* (2017–18) [blog], https://working-worms.net/ (accessed 28 January 2022).

9 I would note, however, that I recoiled from Darwin's rough handling of the worms, in his scientific experiments and 'observations'. See Charles Darwin, *The Formation of Vegetable Mould through the Action of Worms, with Observations on their Habits* (1881; Horcott, Glos: Echo Library, 2007), 15.

10 Gabriel Rockhill, 'Translator's Introduction: Jacques Rancière's Politics of Perception' in *The Politics of Aesthetics: The Distribution of the Sensible*, ed. Jacques Rancière, trans. Gabriel Rockhill (2000; London and New York: Continuum, 2004), 1.

11 Jacques Rancière, *The Politics of Aesthetics: The Distribution of the Sensible*, trans. Gabriel Rockhill (2000; London and New York: Continuum, 2004), 10, 12–19.

12 Critical animal studies scholar Fiona Probyn-Rapsey has pointed to the power of denial in our relations to the exploitation, commodification and instrumentalization of animals – what we see, what we 'know' but what we deny. See Fiona Probyn-Rapsey, 'Human Rights and Animal Ethics Research Network Reading Group' [Zoom Meeting], 23 February 2021.

13 Neumark and Miranda, *Waiting*.

14 Giles Deleuze and Felix Guattari, *A Thousand Plateaus: Capitalism and Schizophrenia*, trans. Brian Massumi (1980; Minneapolis and London: University of Minnesota Press, 1987), 273.

15 Christof Cox, 'Of Humans, Animals, and Monsters', in *Becoming Animal: Contemporary Art in the Animal Kingdom*, ed. Nato Thompson with Christof Cox and Joseph Thompson (North Adams, MA: MASS MoCA Publications, 2005), 23.

16 Donna Haraway, *When Species Meet* (Minneapolis: University of Minnesota Press, 2008), 29.

17 Kathleen Stewart, *Ordinary Affects* (Durham, NC: Duke University Press, 2007), 1, 4, 128.

18 See, for instance, the important work of Lori Gruen and the post-entanglement work of Eva Haifa Giraud: Lori Gruen, *Entangled Empathy: An Alternative Ethic for our Relationships with Animals* (New York: Lantern Books, 2015); Eva Haifa Giraud, *What Comes After Entanglement: Activism, Anthropocentrism, and an Ethics of Exclusion* (Durham, NC and London: Duke University Press, 2019).

19 Paula Arcari, Fiona Probyn-Rapsey, and Hayley Singer, 'Where Species Don't Meet: Invisibilized Animals, Urban Nature and City Limits', *Environment and Planning E: Nature and Space* 4, no. 3 (2021): 940.

20 Meijer, *When Animals Speak*, 154.

21 Stewart, *Ordinary Affects*, 128.

22 ABC RN, 'The Unsung Hero' (2019) [radio programme], https://vimeo.com /426488760 (accessed 28 January 2022).

23 The political and ethical question of zoos is too complex to address here, but there is ample animal studies and ethical environmental literature on the subject. For instance, Matthew Chrulew, 'Managing Love and Death at the Zoo: The Biopolitics of Endangered Species Preservation', in *Unloved Others: Death of the Disregarded in the Time of Extinctions*, ed. Deborah Bird Rose and Thom van Dooren. *Australian Humanities Review* 50 (May 2011): 137–57.

24 Mick Smith, 'Dis(appearance): Earth, Ethics and Apparently', in *Unloved Others: Death of the Disregarded in the Time of Extinctions*, ed. Deborah Bird Rose and Thom van Dooren, *Australian Humanities Review* 50 (May 2011): 35.

25 Smith, 'Dis(appearance)', 35.

26 Deborah Bird Rose and Thom van Dooren, 'Introduction', in *Unloved Others: Death of the Disregarded in the Time of Extinctions*, ed. Deborah Bird Rose and Thom van Dooren, *Australian Humanities Review* 50 (May 2011): 2.

27 Tara Mehrabi, 'Being Intimate with Flies: On Affective Methodologies and Laboratory Work', *Women, Gender & Research* 1 (2018): 75.

28 Mehrabi, 'Being Intimate with Flies', 75; my emphasis.

29 Mehrabi, 'Being Intimate with Flies', 74. Mehrabi cites Haraway, *When Species Meet*, n.p., and provides a quotation from Jacob Bull, 'Between Ticks and People: Responding to Nearbys and Contentments', *Emotion, Space and Society* 12 (2014): n.p. (73–84).

30 Sara Ahmed, *The Cultural Politics of Emotion* (Edinburgh: Edinburgh University Press, 2014), 83–5.

31 Ahmed, *The Cultural Politics of Emotion*, 94.

32 Tara Mehrabi, 'Queer Ecologies of Death in the Lab: Rethinking Waste, Decomposition and Death through a Queerfeminist Lens', *Australian Feminist Studies* 35, no. 104 (2020): 147.

33 Stewart, *Ordinary Affects*, 19.

34 Maria Miranda and Norie Neumark, *It's Still Life: Listening to Compost* [installation], Living Museum of the West, Melbourne, Australia, 2019.

35 Norie Neumark, *Separation Anxiety: Not the Truth About Alchemy* (1997) [radiophonic work], http://www.somewhere.org/ (accessed 28 January 2022).

36 Neumark, *Separation Anxiety*.

37 Jonathan Crary, *Techniques of the Observer: On Vision and Modernity in the Nineteenth Century* (1990; Cambridge, MA and London: MIT Press, 1992), 19.

38 Crary, *Techniques of the Observer*, 17, 19.

39 Probyn-Rapsey, 'Human Rights and Animal Ethics Research Network Reading Group', [Zoom Meeting], 23 February 2021.

40 Tara Mehrabi describes the 'knowledge-production practices through which transgenic fruit flies, in this case are made invisible and banished into a land of nonhumanness, nonbelonging and nongrievability . . . animal models whose bodies are heavily colonized, regulated within the regimes of knowledge production and made killable'. See Tara Mehrabi, *Making Death Matter: A Feminist Technoscience Study of Alzheimer's Sciences in the Laboratory* (Linköping: Linköping University Press, 2016), 49.
41 Jane Bennett, *The Enchantment of Modern Life: Attachments, Crossings and Ethics* (Princeton: Princeton University Press, 2001), 37.
42 Neumark, *Separation Anxiety*; Bennett, *The Enchantment of Modern Life*, 36.
43 Bennett, *The Enchantment of Modern Life*, xi, 34.
44 Bennett, *The Enchantment of Modern Life*, 34.

Bibliography

Abé, Ryūichi. *The Weaving of Mantra: Kūkai and the Construction of Esoteric Buddhist Discourse*. New York: Columbia University Press, 1999.

Acheson, Kris. 'Silence as Gesture: Rethinking the Nature of Communicative Silences'. *Communication Theory* 18, no. 4 (29 October 2008): 535–55.

Acheson, Kris. 'Fences, Weapons, Gifts: Silences in the Context of Addiction'. In *Silence, Feminism, Power: Reflections at the Edges of Sound*, edited by Aimee Carrillo Rowe and Sheena Malhotra, 188–99. New York: Palgrave Macmillan, 2013.

Ahmed, Sara. *The Cultural Politics of Emotion*. Edinburgh: Edinburgh University Press, 2014.

Altman, Rick. 'Moving Lips: Cinema as Ventriloquism'. *Yale French Studies* 60 Cinema/Sound (1980): 67–79.

Altman, Rick. 'The Material Heterogeneity of Recorded Sound'. In *Sound Theory, Sound Practice*, edited by Rick Altman, 15–31. New York and London: Routledge, 1992.

Andress, Sarah. 'Musical Cards: Christian Marclay's *Shuffle*'. *Art on Paper* 12, no. 2 (2007): 24, 26.

Arai, Kōken. 'Shōjitsushō ni tsuite: *Gyosan taigaishū* to no hikaku o chūshin ni'. *Bukkyō bunka gakkai kiyo* 24 (November 2015): 103–40.

Arcari, Paula, Fiona Probyn-Rapsey and Hayley Singer. 'Where Species Don't Meet: Invisibilized Animals, Urban Nature and City Limits'. *Environment and Planning E: Nature and Space* 4, no. 3 (2021): 940–65.

Ashihara, Jakushō, ed. *Gyosan taigaishū*. Wakayama-ken: Kimura Tomematsu, 1892.

Ashley, Robert. 'The Wolfman for Amplified Voice and Tape'. In *Source: Music of the Avant Garde 1966–73*, edited by Larry Austin and Douglas Kahn, 143–5. Berkeley: University of California Press, 2011.

Ashley, Robert. 'Performances'. *robertashley.org*, http://www.robertashley.org/performances/performance-history.htm, accessed 9 February 2021.

Aspin, Clive. 'Hōkakatanga – Māori Sexualities'. https://teara.govt.nz/en/hokakatanga- maori-sexualities/print, accessed 12 December 2021.

Augoyard, Jean-François, and Henry Torgue, eds. *Sonic Experience: A Guide to Everyday Sounds*. Montreal: McGill-Queens University Press, 2005.

Aumont, Jacques. 'Griffith – the Frame, the Figure'. In translated by Judith Ayling and Thomas Elaesser. *Early Cinema: Space, Frame, Narrative*, edited by Thomas Elsaesser with Adam Barker, 348–59. London: BFI Publishing, 1990.

Backović, Vera. *Džentrifikacija kao socio-prostorni fenomen savremenog grada: Sociološka analiza koncepta*. Beograd: Filozofski fakultet, Univerzitet u Beogradu, 2015.

Bailey, Derek. *Musical Improvisation*. Englewood Cliffs, NJ: Prentice Hall, 1980.

Bailey, Matt. 'Noé, Gaspar'. *Senses of Cinema* October, no. 28 (2003), https://www.sensesofcinema.com/2003/great-directors/noe/, accessed 9 November 2020.

Bain, Mark. 'Transient Architecture: On the Making of the Live Room'. *Thresholds* 16, no. Spring (1998): 50–3.

Balance, Christine Bacareza. *Tropical Renditions: Making Musical Scenes in Filipino America*. Durham, NC: Duke University Press, 2016.

Balázs, Béla. *Theory of the Film: Character and Growth of a New Art*. 1952. Translated by Edith Bone. New York: Dover, 1970.

Banerjee, Mita. *Ethnic Ventriloquism: Literary Minstrelsy in Nineteenth-Century American Literature*. Heidelberg: Winter, 2008.

Baron, Jaimie, Jennifer Fleeger, and Shannon Wong Lerner. 'Introduction: Theorizing Media Ventriloquism'. In *Media Ventriloquism: How Audiovisual Technologies Transform the Voice-Body Relationship*, edited by Jaimie Baron, Jennifer Fleeger and Shannon Wong Lerner, 1–17. New York: Oxford University Press, 2021.

Barthes, Roland. *Camera Lucida: Reflections on Photography*. 1980. Translated by Richard Howard. London: Vintage, 2000.

Baugh, Bruce. 'Body'. In *The Deleuze Dictionary*, edited by Adrian Parr, 35–7. Rev. edn. Edinburgh: Edinburgh University Press, 2010.

Baugh, Bruce. 'Death'. In *The Deleuze Dictionary*, edited by Adrian Parr, 63–5. Rev. edn. Edinburgh: Edinburgh University Press, 2010.

Bazin, André. *What is Cinema? Volume 1*. 1967. Translated by Hugh Gray. Berkeley, Los Angeles and London: University of California Press, 2004.

BBC News. 'Cannes Film Sickens Audience'. 26 May 2002, http://news.bbc.co.uk/1/hi/entertainment/2008796.stm, accessed 1 November 2021.

Beatson, Peter. 'Richard Nunns: The Renaissance of Traditional Māori Music'. *Music in the Air* 16 (2003): 17–33.

Beaumont-Thomas, Ben. 'Low: *Double Negative* review – the sound of the world unravelling'. *The Guardian*, 14 September 2018, https://theguardian.com/music/2018/sep/14/low-double-negative-review-the-sound- of-the-world-unravelling, accessed 1 March 2021.

Bellman, Jonathan. 'Indian Resonances in the British Invasion, 1965–1968'. *The Journal of Musicology* 15, no. 1 (1997): 116–36.

Bennett, Jane. *The Enchantment of Modern Life: Attachments, Crossings and Ethics*. Princeton: Princeton University Press, 2001.

Bennett, Jane. *Vibrant Matter*. Durham, NC: Duke University Press, 2010.

Beoinfo. 'Beograd propisuje akustičke zone i dozvoljen nivo buke u njima'. 2020, http://rs.n1info.com/Vesti/a630869/Beograd-propisuje-akusticke-zone-i-dozvoljen- nivo-buke-u-njima.html, accessed 8 February 2021.

Bergson, Henri. *Matter and Memory*. 1896. Translated by Nancy Margaret Paul and W. Scott Palmer. New York: Zone Books, 1991.

Berlant, Lauren. 'Showing Up to Withhold: Pope.L's Deadpan Aesthetic'. In *Showing Up to Withhold*, edited by William Pope.L and Karen Reimer, 107–35. Chicago and London: University of Chicago Press, 2013.

Berlant, Lauren. 'Structures of Unfeeling: *Mysterious Skin*'. *International Journal of Politics, Culture, and Society* 28 (2015): 192–213.
Brennan, Teresa. *The Transmission of Affect*. Ithaca: Cornell University Press, 2004.
Briers, David. 'Huddersfield Contemporary Music Festival: Christian Marclay'. *Art Monthly* (February 2019): 40.
Brindley, Erica. *Music, Cosmology, and the Politics of Harmony in Early China*. Albany: State University of New York Press, 2012.
Brinkema, Eugenie, with composition by Evan Johnson. 'Critique of Silence'. *Differences: A Journal of Feminist Cultural Studies* 22, nos 2–3 (2011): 211–34.
Brinkema, Eugenie. *The Forms of the Affects*. Durham, NC and London: Duke University Press, 2014.
Brown, Harko. *Ngā taonga tākaro: Māori Sports and Games*. Auckland: Raupo, 2008.
Butler, Judith. *Gender Trouble*. 1990. New York and London: Routledge, 1999.
Cage, John. *Silence: Lectures and Writings*. Middletown, CT: Wesleyan University Press, 1973.
Cardew, Cornelius. 'Towards an Ethic of Improvisation'. In *Treatise Handbook*. London: Edition Peters, 1971, Ubuweb, https://www.ubu.com/papers/cardew_ethics.html, accessed 21 August 2021.
Carroll, Nicole L. 'Everything in Its Place: A Conceptual Framework for Anti-Music'. *Sound Scripts* 6, no. 1 (2019): Article 15.
Cavarero, Adriana. 'The Vocal Body'. *Qui Parle: Critical Humanities and Social Sciences* 21, no. 1 (2012): 71–83.
Chen, Ping. 'Modern Written Chinese in Development'. *Language in Society* 22, no. 4 (1993): 505–37.
Chion, Michel. *The Voice in Cinema*. 1982. Translated by and edited by Claudia Gorbman. New York: Columbia University Press, 1999.
Chion, Michel. *Audio-Vision: Sound on Screen*. 1990. Translated by Claudia Gorbman and with a foreward by Walter Murch. New York: Columbia University Press, 1994.
Chion, Michel. *Film, A Sound Art*. 2003. Translated by Claudia Gorbman. New York: Columbia University Press, 2009.
Chow, Rey. 'Listening After 'Acousmaticity': Notes on a Transdisciplinary Problematic'. In *Sound Objects*, edited by James Steintrager and Rey Chow, 113–29. Durham, NC: Duke University Press, 2019.
Chrulew, Matthew. 'Managing Love and Death at the Zoo: The Biopolitics of Endangered Species Preservation'. In *Unloved Others: Death of the Disregarded in the Time of Extinctions*, edited by Deborah Bird Rose and Thom Van Dooren. *Australian Humanities Review* 50 (May 2011): 137–57.
Clark, Gillian. 'Introduction'. In *Iamblichus: On The Pythagorean Life*. Translated with notes and introduction by Gillian Clark, ix–xxi. Liverpool: Liverpool University Press, 1989.
Clarke, Bruce. *Gaian Systems, Lynn Margulis, Neocybernetics and the End of the Anthropocene*. Minneapolis and London: University of Minnesota Press, 2020.

Clerk, Carol. 'The Making of 'Hong Kong Garden''. *Uncut*, May 2008, https://
www.rocksbackpages.com/Library/Article/the-making-of-hong-kong
-garden, accessed 1 March 2021.

Clifford, James. 'Of Other Peoples: Beyond the 'Salvage' Paradigm'. In
Discussions in Contemporary Culture #1, edited by Hal Foster, 121–30.
Seattle: Bay Press, 1987.

Clough, Patricia T. 'The Affective Turn: Political Economy, Biomedia and
Bodies'. *Theory, Culture & Society* 25, no. 1 (2008): 1–22.

Colman, Felicity J. 'Affect'. In *The Deleuze Dictionary*, edited by Adrian Parr,
11–14. Rev. edn. Edinburgh: Edinburgh University Press, 2010.

Comer, Thomas, and Silas S. Steele. 'Aladdin Quick Step'. In *Favorite Melodies
From the Grand Chinese Spectacle of Aladdin, or, The Wonderful Lamp*.
Boston, MA: Oliver Ditson, 1874, https://levysheetmusic.mse.jhu.edu/
collection/162/005, accessed 28 March 2022.

Comer, Thomas, and Silas S. Steele. 'Come, Come Away'. In *Favorite Melodies
From the Grand Chinese Spectacle of Aladdin, or, The Wonderful Lamp*.
Boston, MA: Oliver Ditson, 1874, https://levysheetmusic.mse.jhu.edu/
collection/123/020, accessed 28 March 2022.

Connor, Steven. *Dumbstruck: A Cultural History of Ventriloquism*. Oxford:
Oxford University Press, 2000.

Corbett, John. *Extended Play: Sounding Off from John Cage to Dr. Funkenstein*.
Durham, NC: Duke University Press, 1994.

Coutinho, Eduardo, Klaus R. Scherer and Nicola Dibben. 'Singing and Emotion'.
In *The Oxford Handbook of Singing*, edited by Graham F. Welch, David M.
Howard and John Nix, 297–314. Oxford: Oxford University Press, 2014.

Cox, Christoph. 'Of Humans, Animals, and Monsters'. In *Becoming Animal:
Contemporary Art in the Animal Kingdom*, edited by Nato Thompson with
Christoph Cox and Joseph Thompson, 18–25. North Adams, MA: MASS
MoCA Publications, 2005.

Cox, Christoph. 'Beyond Representation and Signification: Toward a Sonic
Materialism'. *Journal of Visual Culture* 10, no. 2 (2011): 145–61.

Cox, Christoph. 'Sonic Realism and Auditory Culture: A Reply to Marie
Thompson and Annie Goh'. *Parallax* 24, no. 2 (2018): 234–42.

Cox, Christoph. *Sonic Flux: Sound, Art, and Metaphysics*. Chicago: The
University of Chicago Press, 2018.

Crary, Jonathan. *Techniques of the Observer: On Vision and Modernity in the
19th Century*. 1990. Cambridge, MA and London: The MIT Press, 1992.

Dahlhaus, Carl. *Nineteenth-Century Music*. Translated by J. Bradford Robinson.
Berkeley: University of California Press, 1989.

Darwin, Charles. *The Formation of Vegetable Mould through the Action of Worms,
with Observations on their Habits*. 1881. Horcott, Glos: Echo Library, 2007.

Davies, David. 'Watching the Unwatchable: *Irréversible*, Empire, and the
Paradox of Intentionally Inaccessible Art'. In *Suffering Art Gladly: The
Paradox of Negative Emotion in Art*, edited by Jerrold Levinson, 246–65.
Basingstoke and New York: Palgrave Macmillan, 2014.

Debord, Guy. 'Introduction to a Critique of Urban Geography'. In *Situationist International Anthology*, edited and translated by Ken Knabb, 5–8. Berkeley: Bureau of Public Secrets, 1981.

Deleuze, Gilles, *Difference and Repetition*. 1968. Translated by Paul Patton. London: Athlone Press, 1994.

Deleuze, Gilles. *The Logic of Sense*. 1969. Translated by Mark Lester. London: The Athlone Press, 1990.

Deleuze, Gilles. *Spinoza: Practical Philosophy*. 1970. Translated by Robert Hurley. San Francisco: City Lights Books, 1988.

Deleuze, Gilles. *Francis Bacon: The Logic of Sensation*. 1981. Translated by with an introduction by Daniel W. Smith and an afterword by Tom Conley. Minneapolis: University of Minnesota Press, 2003.

Deleuze, Gilles. *Cinema 1: The Movement-Image*. 1983. Translated by Hugh Tomlinson and Barbara Habberjam. Minneapolis: University of Minnesota Press, 1986.

Deleuze, Gilles. *Cinema 2: The Time-Image*. 1985. Translated by Hugh Tomlinson and Robert Galeta. Minneapolis: University of Minnesota Press, 1989.

Deleuze, Gilles. *The Fold: Leibniz and the Baroque*. 1988. Translated by Tom Conley. London: The Athlone Press, 1993.

Deleuze, Gilles, and Félix Guattari. *A Thousand Plateaus: Capitalism and Schizophrenia*. 1980. Translated by Brian Massumi. Minneapolis and London: University of Minnesota Press, 1987.

Derrida, Jacques. *Positions*. 1972. Translated by Alan Bass. Chicago: University of Chicago Press, 1981.

Derrida, Jacques. *Specters of Marx: The State of the Debt, the Work of Mourning and the New International*. 1993. Translated by Peggy Kamuf. New York and London: Routledge, 2006.

Derrida, Jacques. *The Animal That Therefore I Am*. 2002. Translated by D. Wills. New York: Fordham University Press, 2008.

Despret, Vinciane. 'The Body We Care For: Figures of Anthropo-zoo-genesis'. *Body and Society* 10, nos 2–3 (2004): 125–70.

Despret, Vinciane. 'Responding Bodies and Partial Affinities in Human-Animal Worlds'. *Theory, Culture and Society* 30, no. 7/8 (2013): 51–76.

Despret, Vinciane. *What Animals Would Say if We Asked the Right Question*. Minneapolis: University of Minnesota Press, 2016.

Deusner, Stephen 'Low – *Double Negative*'. *Uncut*, 9 October 2018, https://www.uncut.co.uk/reviews/low-double-negative-107640/, accessed 1 March 2021.

DeWoskin, Kenneth J. *A Song for One or Two: Music and the Concept of Art in Early China*. Ann Arbor: University of Michigan Press, 1982.

DeWoskin, Kenneth J. 'Philosophers on Music in Early China'. *The World of Music* 27, no. 1 (1985): 33–47.

Dickson, W. K. L. and Antonia Dickson. *History of the Kinetograph, Kinetoscope, and Kineto-phonograph*. New York: Museum of Modern Art, 2001.

Didi-Huberman, Georges. *Invention of Hysteria: Charcot and the Photographic Iconography of the Salpêtrière*. 1982. Translated by Alisa Hartz. Cambridge, MA and London: The MIT Press, 2003.

Didi-Huberman, Georges. 'The Index of the Absent Wound (Monograph on a Stain)'. October 29 (Summer 1984): 63–81.

Doane, Mary Ann. *The Desire to Desire: The Woman's Film of the 1940s*. Bloomington and Indianapolis: Indiana University Press, 1987.

documenta 14. 'Pope.L'. https://www.documenta14.de/en/artists/13513/pope-l, accessed 4 May 2020.

Dolar, Mladen. *A Voice and Nothing More*. Cambridge, MA: The MIT Press, 2006.

Drake, Cathryn. 'Here's Why Greece Is Not Exactly Rolling Out the Red Carpet for documenta 14'. *artnet news*, 29 March 2017, https://news.artnet.com/art-world/greece-not-rolling-out-red-carpet-documenta-14-905459, accessed 21 May 2020.

Drew, Philip. *Leaves of Iron*. Sydney: Law Book Company, 1985.

Durham, Meenakshi Gigi. 'Displaced Persons: Symbols of South Asian Femininity and the Returned Gaze in U.S. Media Culture'. *Communication Theory* 11, no. 2 (2001): 201–17.

Durie, Mason. *Mauri Ora: The Dynamics of Māori Health*. Auckland: Oxford University Press, 2001.

Dyer, Richard. *Only Entertainment*. New York and London: Routledge, 1992.

Eidsheim, Nina S., and Katherine Meizel. 'Introduction'. In *The Oxford Handbook of Voice Studies*, edited by Nina S. Eidsheim and Katherine Meizel, xiii–xxxix. Oxford: Oxford University Press, 2019.

Eisenlohr, Patrick. *Sounding Islam: Voice, Media, and Sonic Atmospheres in an Indian Ocean World*. Berkeley: University of California Press, 2018.

Eisenstein, Sergei. *S. M. Eisenstein: Selected Works. Vol. 1 Writings, 1922–34*. Edited and translated by Richard Taylor. London: BFI Publishing, 1988.

English, Darby. *How to See a Work of Art in Total Darkness*. Cambridge, MA: The MIT Press, 2007.

English, Darby. *To Describe a Life: Notes on the Intersection of Art and Race Terror*. New Haven and London: Yale University Press, 2019.

Ernst, Douglas. 'Michael Eric Dyson: Kanye West's Support of Trump is "White Supremacy by Ventriloquism"'. *The Washington Times*, 12 October 2018, https://www.washingtontimes.com/news/2018/oct/12/michael-eric-dyson-kanye-wests-support-of-trump-is/, accessed 9 December 2021.

Eubanks, Charlotte. 'Sympathetic Response: Verbal Arts and the Erotics of Persuasion in the Buddhist Literature of Medieval Japan'. *Harvard Journal of Asiatic Studies* 72, no. 1 (2012): 43–70.

Eubanks, Charlotte. 'Cultures of Sound: Lineages and Languages of Sutra Recitation in Goshirakawa's Japan'. In *The Languages of Religion*, edited by Sipra Mukherjee, 17–34. London: Routledge, 2018.

Everest, Alton F. *The Master Handbook of Acoustics*, 4th edn. New York: McGraw-Hill, 2001.

Farrell, B. A. 'Catharsis'. In *The New Fontana Dictionary of Modern Thought*. 1977, edited by Alan Bullock and Stephen Trombley. 4th edn. London: Harper Collins, 1999.

Fast, Susan. *In the Houses of the Holy: Led Zeppelin and the Power of Rock Music*. Oxford: Oxford University Press, 2001.

Ferrett, D. 'The Black Hole Song of Unsounding Mothers'. In *Dark Sound: Feminine Voices in Sonic Shadow*, 107–44. New York: Bloomsbury Academic, 2020.

Filipović, Andrija. 'Resonant Masculinities: Affective Co-production of Sound, Space, and Gender in the Everyday Life of Belgrade, Serbia'. *NORMA: International Journal for Masculinity Studies* 13, nos 3–4 (2019): 213–26.

Flintloff, Brian. *Taonga Pūoro, Singing Treasures: The Musical Instruments of the Māori* Nelson: Potter & Burton, 2004.

Fournier, Felix Alfaro. *Playing Cards*. Vitoria: Fournier Museum, 1982.

Fowkes, Maja. *The Green Bloc: Neo-avant-garde Art and Ecology under Socialism*. Budapest: Central European University Press, 2015.

Fowler, Leo. *Te Mana o Turanga: The Story of the Carved House Te Mana o Turanga on the Whakato Marae at Manutuke Gisborne*. Auckland: N.Z. Historic Places Trust, 1974.

Freud, Sigmund. 'Frau Emmy von N, Case Histories from Studies on Hysteria'. In Josef Breuer and Sigmund Freud's *The Standard Edition of the Complete Psychological Works of Sigmund Freud*, Volume II (1893-1895): *Studies on Hysteria*, 1893, translated and edited by James Strachey, 48–105. London: Hogarth Press, 1955.

Frith, Simon. *Performing Rites: On the Value of Popular Music*. Cambridge, MA: Harvard University Press, 1996.

Gann, Kyle. *No Such Thing as Silence: John Cage's 4'33"*. New Haven: Yale University Press, 2010.

Garbutt, Rob. *The Locals*. Bern: Peter Lang, 2011.

Garrett, Charles Hiroshi. 'Chinatown, Whose Chinatown? Defining America's Borders with Musical Orientalism'. *Journal of the American Musicological Society* 57, no. 1 (2004): 119.

Gautier, Ana María Ochoa. 'Silence'. In *Keywords in Sound*, edited by David Novak and Matt Sakakeeny, 183–207. Durham, NC and London: Duke University Press, 2015.

Gavrilović, Dušan. *Opštine i regioni u Republici Srbiji*. Beograd: Republički zavod za statistiku, 2020.

Gayraud, Agnès. 'Glossolalia / Xenoglossia'. In *Unsound: Undead*, edited by Steve Goodman, Toby Heys and Eleni Ikoniadou, 61–4. Falmouth: Urbanomic, 2019.

Giraud, Eva Haifa. *What Comes After Entanglement: Activism, Anthropocentrism, and an Ethics of Exclusion*. Durham, NC and London: Duke University Press, 2019.

Goad, Philip. *Troppo*. Pesaro Architectural Monographs. Edited by Patrick Bingham-Hall. Balmain: Pesaro Publishing, 1999.

Goh, Annie. 'Sounding Situated Knowledges: Echo in Archaeoacoustics'. *Parallax* 23, no. 3 (July 3, 2017): 283–304.

Goldsmith, Mike. '*Double Negative*, Low'. *Record Collector Magazine*, 7 August 2018, https://recordcollectormag.com/reviews/album/double-negative, accessed 1 March 2021.

Goodman, Steve. *Sonic Warfare: Sound, Affect, and the Ecology of Fear.* Cambridge, MA and London: The MIT Press, 2012.

Gradski zavod za javno zdravlje Beograd. 'Rezultati merenja buke u životnoj sredini za jesenji ciklus 2020'. godine: Podaci o kvalitetu – oktobar 2020, https://www.beograd.rs/lat/gradska-vlast/1746715-podaci-o-kvalitetu -cinilaca- zivotne-sredine/, accessed 8 Feb. 2021.

Graves, Kirk Walker. *Kanye West's My Beautiful Dark Twisted Fantasy.* London: Bloomsbury, 2014.

Grossberg, Lawrence. *We Gotta Get Out of This Place: Popular Conservatism and Postmodern Culture.* New York and London: Routledge, 1992.

Gruen, Lori. *Entangled Empathy: An Alternative Ethic for our Relationships with Animals.* New York: Lantern Books, 2015.

Gunning, Tom. 'The Cinema of Attractions: Early Film, Its Spectator, and the Avant-Garde'. In *The Cinema of Attractions Reloaded*, edited by Wanda Strauven, 381–8. Amsterdam: Amsterdam University Press, 2006.

Gunning, Tom. 'Weaving a Narrative: Style and Economic Background in Griffith's Biograph Films'. In *Early Cinema: Space, Frame, Narrative*, edited by Thomas Elsaesser with Adam Barker, 336–47. London: BFI Publishing, 1981.

Hakeda, Yoshito S. *Kūkai: Major Works.* New York: Columbia University Press, 1972.

Haraway, Donna. *When Species Meet.* Minneapolis: University of Minnesota Press, 2008.

Harrison, Bertha. 'Games of Music'. *The Musical Times* 48, no. 775 (1907): 589–92.

Haumanu Collective. 'Haumanu Collective'. https://www.haumanucollective .com/, accessed 10 December 2021.

Heller-Nicholas, Alexandra. *Rape-Revenge Films: A Critical Study.* Jefferson, North Carolina; London: McFarland, 2011.

Hemmings, Clare. 'Affective Solidarity: Feminist Reflexivity and Political Transformation'. *Feminist Theory* 13, no. 2 (August 2012): 147–61.

Henare, Manuka. 'Tapu, Mana, Mauri, Hai, Wairua: A Māori Philosophy of Vitalism and Cosmos'. In *Indigenous Traditions and Ecology: The Interbeing of Cosmology and Community*, edited by J. Grimm, 197–221. Cambridge, MA: Harvard University Press, 2001.

Hisama, Ellie M. 'Postcolonialism on the Make: The Music of John Mellencamp, David Bowie and John Zorn'. *Popular Music* 12, no. 2 (1993): 91–104.

Hogan, Jackie. 'Staging the Nation'. *Journal of Sport and Social Issues* 27, no. 2 (2003): 100–23.

Holmes, Thom. *Electronic and Experimental Music: Technology, Music and Culture*. New York: Routledge, 2016.

Holterbach, Emmanuel. 'Eliane Radigue: Feedback Works 1969–1970'. In the liner notes for Eliane Radigue, *Feedback Works 1969–70* (2014) [audio recording], Italy: Alga Marghen.

Holterbach, Emmanuel. 'Eliane Radigue: Opus 17'. in the liner notes to Eliane Radigue, *Opus 17* (2015) [audio recording], Italy: Alga Marghen.

hooks, bell. 'Sisterhood: Political Solidarity between Women'. *Feminist Review* 23, no. 23 (1986): 125–38.

hooks, bell. *The Will to Change: Men, Masculinity, and Love*. New York: Washington Square Press, 2004.

Horn, Katrin. *Women, Camp, and Popular Culture: Serious Excess*. Cham: Palgrave MacMillan, 2017.

Hoskins, Te Kawehau, and Alison Jones. 'Non-Human Others and Kaupapa Māori Research'. 2017. In *Critical Conversations in Kaupapa Māori*, edited by Te Kawehau Hoskins and Alison Jones, 49–64. Wellington: Huia Publishers, 2021.

Hove, Michael J., Peter Vuust and Jan Stupacher. 'Increased Levels of Bass in Popular Music Recordings 1955–2016 and Their Relation to Loudness'. *The Journal of the Acoustical Society of America* 145, no. 4 (2019: 2247–53).

Hughes-d'Aeth, Tony. 'Kim Scott's Taboo and the Extimacy of Massacre'. *Journal of Australian Studies* 45, no. 2 (2021): 165–80.

Hurvitz, Leon, trans. *Scripture of the Lotus Blossom of the Fine Dharma, Translated from the Chinese of Kumārajīva*. New York: Columbia University Press, 1976.

Ireland, Brian, and Sharif Gemie. 'Raga Rock: Popular Music and the Turn to the East in the 1960s'. *Journal of American Studies* 53, no. 1 (2019): 57–94.

Iyer, Vijay. 'On Improvisation, Temporality, and Embodied Experience'. In *Sound Unbound: Sampling Digital Music and Culture*, edited by Paul D. Miller, 273–92. Cambridge, MA: The MIT Press, 2008.

Jamieson, Nigel. 'Make It Relevant'. *Australasian Drama Studies* 54 (2009): 105–24.

Jasen, Paul C. *Low End Theory: Bass, Bodies and the Materiality of Sonic Experience*. New York; London; Oxford; New Delhi; Sydney: Bloomsbury Academic, 2016.

Johnson, Julia R. 'Qwe're Performances of Silence: Many ways to Live "Out Loud."' In *Silence, Feminism, Power: Reflections at the Edges of Sound*, edited by Aimee Carrillo Rowe and Sheena Malhotra, 50–66. New York: Palgrave Macmillan, 2013.

Jones, Alison. *This Pākehā Life: An Unsettled Memoir*. Wellington: Bridget William Books, 2020.

Juzwiak, Rich. 'Low, *Double Negative*'. *Pitchfork*, 14 September 2018, https://pitchfork.com/reviews/albums/low-double-negative/, accessed 1 March 2021.

Kafka, Franz. *The Metamorphosis*. 1915. Translated, edited and with an introduction by Stanley Corngold. New York: Modern Library New York, 1972.

Kahane, Claire. *Passions of the Voice: Hysteria, Narrative and the Figure of the Speaking Woman 1850–1915*. Baltimore and London: The John Hopkins University Press, 1995.

Kane, Brian. *Sound Unseen*. New York: Oxford University Press, 2014.

Kapchan, Deborah. 'Body'. In *Keywords in Sound*, edited by David Novak and Matt Sakakeeny, 33–44. Durham, NC and London: Duke University Press, 2015.

Kawaguchi, Hisao and Nobuyoshi Shida, eds. *Wakan rōeishū, Ryōjin hishō*. Nihon Koten Bungaku Taikei, vol. 73. Tokyo: Iwanami Shoten, 1965.

Keller, William B. *A Catalogue of the Cary Collection of Playing Cards in the Yale University Library*, Vol. I and II. New Haven: Yale University Library, 1981.

Kim, Yung-Hee. *Songs to Make the Dust Dance: The Ryojin Hisho of Twelfth-Century Japan*. Berkeley: University of California Press, 1994.

Komene, Jo'el. 'Kōauau auē, e auau tō au e! The Kōauau in Te Ao Māori'. MA diss., University of Waikato, Hamilton, New Zealand, 2009.

Kontturi, Katve-Kaisa. *Ways of Following: Art, Materiality, Collaboration*. London: Open Humanities Press, 2018.

Kristeva, Julia. *Powers of Horror: An Essay on Abjection*. 1980. Translated by Leon S. Roudiez. New York: Columbia University Press, 1982.

Kruger, Tāmati. 'The Qualities of Ihi, Wehi and Wana'. In *Nga Tikanga Tuku Iho a te Māori, Customary Concepts of the Māori: A Source Book for Māori Studies Students*, edited by Hirini Moko Mead, 228–45. Wellington: Victoria University of Wellington, 1984.

LaBelle, Brandon. *Lexicon of the Mouth: Poetics and Politics of Voice and the Oral Imaginary*. New York: Bloomsbury, 2014.

Lacan, Jacques. *Écrits*. 1977. London: Tavistock, 1982.

Lacan, Jacques. *The Ethics of Psychoanalysis*. 1986. Translated by Dennis Porter. London: Routledge, 2008.

Lancefield, Robert Charles. 'Hearing Orientality in (White) America, 1900–1930'. PhD diss., Wesleyan University, 2004.

'Larynx'. *Encyclopædia Britannica*, https://www.britannica.com/science/larynx, accessed 11 December 2021.

Lawrence, David Herbert. *Kangaroo*. 1922. London: Heinemann, 1970.

Leas, Ryan. 'Album of the Week: Low *Double Negative*'. *Stereogum*, 11 September 2018, https://www.stereogum.com/2013136/low-double-negative-review/reviews/album-of- the-week/, accessed 1 March 2021.

Lennon, John and Ono, Yoko. Interview with David Sheff. *Playboy*, January 1981.

Lerner, Neil, ed. *Music in the Horror Film: Listening to Fear*. New York and London: Routledge, 2010.

Leventhall, Geoff. 'Low Frequency Noise: What We Know, What We Do Not Know, and What We Would like to Know'. *Journal of Low Frequency Noise, Vibration and Active Control* 28, no. 2 (1 June 2009): 79–104.

Levitt, Rachel. 'Silence Speaks Volumes: Counter-Hegemonic Silences, Deafness, and Alliance Work'. In *Silence, Feminism, Power: Reflections at the Edges of Sound*, edited by Aimee Carrillo Rowe and Sheena Malhotra, 67–82. New York: Palgrave Macmillan, 2013.

Lipsitz, George. *Dangerous Crossroads: Popular Music, Postmodernism, and the Poetics of Place*. London: Verso, 1994.

Lorde, Audre. *The Collected Poems of Audre Lorde*. New York and London: W. W. Norton & Company, 2017.

Lott, Eric. *Love & Theft: Blackface Minstrelsy and the American Working Class*. 1993. Oxford: Oxford University Press, 2013.

Lucier, Alvin. 'My Affairs with Feedback'. *Resonance* 9, no. 2 (2002): 24–5.

Lukowski, Andrzej. 'Low, *Double Negative*'. *Drowned in Sound*, 13 September 2018, https://drownedinsound.com/releases/20433/reviews/4152016, accessed 1 March 2021.

Lyotard, Jean-François. 'Several Silences'. 1972. In *Driftworks*, edited by Roger McKeon and translated by Joseph Maier, 91–110. New York: Semiotext(e), 1984.

Malabou, Catherine. 'La plasticité en souffrance'. *Sociétés & Représentations* 20 (2005): 31–9.

Malabou, Catherine. *Plasticity at the Dusk of Writing: Dialectic, Destruction, Deconstruction*. 2005. Translated by Carolyn Shread. New York: Columbia University Press, 2010.

Malabou, Catherine. *The New Wounded: From Neurosis to Brain Damage*. 2007. Translated by Steven Miller. New York: Fordham University Press, 2012.

Malhotra, Sheena, and Aimee Carrillo Rowe. *Silence, Feminism, Power: Reflections at the Edges of Sound*. Basingstoke: Palgrave Macmillan, 2013.

Manning, Erin, *Relationscapes: Movement, Art, Philosophy*. Cambridge, MA: The MIT Press, 2009.

Massumi, Brian. *Parables for the Virtual: Movement, Affect, Sensation*. Durham, NC and London: Duke University Press, 2002.

Massumi, Brian. 'Notes of the Translation and Acknowledgement'. In *A Thousand Plateaus: Capitalism and Schizophrenia*, edited by Gilles Deleuze and Felix Guattari. 1980, translated by Brian Massumi. London and New York: Continuum, 2004.

Massumi, Brian. *Politics of Affect*. Cambridge: Polity Press, 2015.

Massumi, Brian. *The Principle of Unrest: Activist Philosophy in the Expanded Field*. London: Open Humanities Press, 2017, http://openhumanitiespress .org/books/titles/the-principle-of-unrest, accessed 1 March 2021.

Maturana, Humberto R. 'Biology of Cognition'. In *Autopoiesis and Cognition: The Realization of the Living*, edited by Humberto R. Maturana and Francisco J. Varela, 1–58. Dordrecht: D. Reidel Publishing Company, 1980.

McClary, Susan. *Conventional Wisdom: The Content of Musical Form*. Berkeley: University of California Press, 2000.

McClary, Susan. *Feminine Endings: Music, Gender & Sexuality*. 1991. Minneapolis and London: University of Minnesota Press, 2002.

McLaren, Annemarie. 'A Many-Sided Frontier'. *Australian Historical Studies* 50, no. 2 (2019): 235–54.

McNeill, Hinematau, and Sandy Hata. 'Karanga'. In *Kia Rōnaki: The Māori Performing Arts*, edited by Rachael Ka'ai-Mahuta, Tania Ka'ai and John Moorfield, 53–60. Auckland: Pearson, 2013.

Mead, Hirini Moko. *Tikanga Māori: Living By Māori Values*. Rev. edn. Wellington: Huia Publishing, 2016.

Mehrabi, Tara. *Making Death Matter: A Feminist Technoscience Study of Alzheimer's Sciences in the Laboratory*. Linköping: Linköping University Press, 2016.

Mehrabi, Tara. 'Being Intimate with Flies: On Affective Methodologies and Laboratory Work'. *Women, Gender & Research* 1 (2018): 73–80.

Mehrabi, Tara. 'Queer Ecologies of Death in the Lab: Rethinking Waste, Decomposition and Death through a Queerfeminist Lens'. *Australian Feminist Studies* 35, no. 104 (2020): 138–54.

Meijer, Eva. *When Animals Speak: Toward an Interspecies Democracy*. New York: New York University Press, 2019.

Metzer, David. 'The Power Ballad'. *Popular Music* 31(2012): 437–59.

Mika, Carl Te Hira. 'The Co-Existence of Self and Thing Through Ira: A Māori Phenomenology'. *Journal of Aesthetics and Phenomenology* 2, no. 1 (2015): 93–112.

Mika, Carl Te Hira. 'The Uncertain Kaupapa of Kaupapa Māori'. 2017. In *Critical Conversations in Kaupapa Māori*, edited by Te Kawehau Hoskins and Alison Jones, 119–32. Wellington: Huia Publishers, 2021.

Miles, Stephen. 'Objectivity and Intersubjectivity in Pauline Oliveros's "Sonic Meditations"'. *Perspectives of New Music* 46, no. 1 (2008): 4–38.

Miller, Jacques-Alain. 'Extimité'. In *Lacanian Theory of Discourse*, edited by Mark Bracher, Marshall Alcorn, Ronald Corthell and Françoise Massardier-Kenney, 74–87. New York: New York University Press, 1994.

Mitchell, Juliet. *Women: The Longest Revolution: Essays in Feminism, Literature and Psychoanalysis*. London: Virago Press, 1984.

Mookherjee, Nayanika. 'Mobilising Images: Encounters of 'Forced' Migrants and the Bangladesh War of 1971'. *Mobilities* 6, no. 3 (2011): 399–414.

Moon, Krystyn R. *Yellowface: Creating the Chinese in American Popular Music and Performance, 1850s–1920s*. New Brunswick: Rutgers University Press, 2005.

Moon, Paul. *A Tohunga's Natural World; Plants, Gardening and Food*. Auckland: David Ling Publishing, 2005.

Moorfield, John. *Te Aka Māori Dictionary*, https://maoridictionary.co.nz/, accessed 30 October 2021.

Morgan, Frances. 'Darkness Audible: Sub-Bass, Tape Decay and Lynchian Noise'. In *The End: An Electric Sheep Anthology*, edited by Virginie Sélavy, 186–202. London: Attractor Press, 2011.

Mornement, Adam, and Simon Holloway. *Corrugated Iron*. London: Frances Lincoln, 2007.

Morton, Timothy . 'Appearance is War'. In *The War of Appearances: Transparency, Opacity, Radiance*, edited by Joke Brouwer, Lars Spuybroek and Sjoerd van Tuinen, 166–82. Rotterdam: V2_Publishing, 2016.

Muldoon, Mark S. 'Silence Revisited: Taking the Sight out of Auditory Qualities'. *The Review of Metaphysics* 50, no. 2 (1996): 275–98.

Müller, Leopold. *What the Doctor Overheard: Dr. Leopold Müller's Account of Music in Early Meiji Japan (Einege Notizen über die japanische Musik, 1874–1876)*. Edited and translated by Elizabeth Markham, Naoko Terauchi, and Rembrandt Wolpert. Ithaca: Cornell East Asia Series, 2017.

Murray, Charles Shaar. 'David Bowie: Sermon from The Savoy'. *New Musical Express*, 29 September 1984, http://www.rocksbackpages.com/Library/Article/david-bowie-sermon-from-the-savoy, accessed 28 March 2022.

Musser, Charles. *Before the Nickelodeon: Edwin S. Porter and the Edison Manufacturing Company*. Berkeley, Los Angeles and Oxford: University of California Press, 1991.

Musser, Charles. 'A Cornucopia of Images: Comparison and Judgment across Theatre, Film, and the Visual Arts during the Late Nineteenth Century'. In *Moving Pictures: American Art and Early Film, 1880–1910*, edited by Nancy Mowll Mathews with Charles Musser, 5–38. New York: Hudson Hills Press, 2005.

Myers, Peter. 'Corrugated Galvanised Iron'. *Transition* 1–2 (1981): 24–6.

Nail, Thomas. *Being and Motion*. Oxford: Oxford University Press, 2019.

Neumark, Norie. *Voicetracks: Attuning to Voice in Media and the Arts*. Cambridge, MA: MIT Press, 2017.

Newman, Steve. 'The Value of 'Nothing': Ballads in 'The Beggar's Opera''. *The Eighteenth Century* 45, no. 3 (2004): 265–83.

Nezavisne. 'Zabrana uvoza Euro3 vozila u Srbiji realna do 2021'. 2019, https://www.nezavisne.com/automobili/auto-novosti/Zabrana-uvoza-Euro3-vozila-u- Srbiji-realna-od-2021/568598, accessed 8 February 2021.

Ngai, Sianne. 'Visceral Abstractions'. *GLQ: A Journal of Lesbian and Gay Studies* 21, no. 1 (2015): 33–63.

Normand, Brigitte Le. *Designing Tito's Capital: Urban Planning, Modernism, and Socialism in Belgrade*. Pittsburgh: University of Pittsburgh Press, 2014.

North, Michael. *The Dialect of Modernism: Race, Language, and Twentieth-Century Literature*. Oxford: Oxford University Press, 1998.

Nunns, Richard. *Te Ara Puoro: A Journey into the World of Māori Music*. Nelson: Craig Potton Publishing, 2014.

Nyman, Michael. *Experimental Music: Cage and Beyond*. New York: Cambridge University Press, 1999.

O'Dell, Paul. *Griffith and the Rise of Hollywood*. New York: A. S. Barnes & Co; London: A. Zwemmer Ltd, 1970.

O'Grady, Megan. 'Answering Society's Thorniest Questions, With Performance Art'. *The New York Times Style Magazine*, 2 March 2018, https://www.nytimes.com/2018/03/02/t-magazine/pope-l-artist.html, accessed 21 May 2020.

Okihiro, Gary Y. 'When and Where I Enter'. In *Asian American Studies Now*, edited by Jean Yu-wen Shen Wu and Thomas C. Chen, 3–20. New Brunswick: Rutgers University Press, 2010.

Oliveros, Pauline. *Sonic Meditations: March–November 1971*. Urbana, IL: Smith Publications, 1974.

Oliveros, Pauline. *Deep Listening: A Composer's Sound Practice*. New York: Universe, 2005.

Oliveros, Pauline. 'Auralizing in the Sonosphere: A Vocabulary for Inner Sound and Sounding'. *Journal of Visual Culture* 10, no. 2 (August 2011): 162–8.

Paasonen, Susanna. *Carnal Resonance: Affect and On-line Pornography*. Cambridge, MA and London: MIT Press, 2011.

Palmer, Tim. 'Style and Sensation in the Contemporary French Cinema of the Body'. *Journal of Film and Video* 58, no. 3 (Fall 2006): 22–32.

Palmer, Tim. 'Under Your Skin: Marina de Van and the Contemporary French *cinéma du corps*'. *Studies in French Cinema* 6, no. 3 (2006): 171–81.

Pearis, Bill. 'Low – *Double Negative*'. *Brooklyn Vegan*, 21 December 2018, https://www.brooklynvegan.com/brooklynvegans-top-50-albums-of-2018/, accessed 1 March 2021.

Peirce, Charles Sanders. *Peirce on Signs: Writings on Semiotic by Charles Sanders Peirce*. Chapel Hill and London: The University of North Carolina Press, 1991.

Perlson, Hili. 'The Tao of Szymczyk: documenta 14 Curator Says to Understand His Show, Forget Everything You Know'. *artnet news*, 6 April 2017, https://news.artnet.com/art- world/adam-szymczyk-press-conference-document a-14-916991, accessed 21 May 2020.

Petrović, Jelisaveta, and Vera Backović, eds. *Postsocialist Capitalism: Urban Changes and Challenges in Serbia*. Belgrade: Institute for Sociological Research, 2019.

Petrušić, Nataša. *Kvalitet životne sredine u Beogradu u 2018. godini*. Beograd: Sekretarijat za zaštitu životne sredine, 2018.

Petrušić, Sandra. 'Temeljna betonizacija – Naprednjačka humanizacija grada'. 2018, https://nedavimobeograd.rs/nin-temeljnabetonizacija-naprednjacka -humanizacija- grada, accessed 8 February 2021.

Phillips, Glen. 'February'. *Journal of Australian Studies* 28, no. 83 (2004): 130.

Piekut, Benjamin. *Experimentalism Otherwise: The New York Avant-Garde and Its Limits*. Berkeley: University of California Press, 2011.

Poe, Edgar Allan, and Hervey Allen. 'Silence – A Fable'. In *The Complete Tales and Poems of Edgar Allan Poe*. New York: The Modern Library, 1938.

Poe, Edgar Allan. 'The Cask of Amontillado'. 1846. In *The Fall of the House of Usher and Other Writings: Poems, Tales, Essays and Reviews*, edited by with an introduction and notes by David Galloway, 310–16. London, New York, Camberwell, Toronto, New Dehli, Auckland and Rosebank: Penguin Books, 2003.

Pope.L and Dieter Roelstraete. *Pope.L: Campaign*. Milan: Mousse Publishing, 2019.

Povinelli, Elizabeth A. *Economies of Abandonment: Social Belonging and Endurance in Late Liberalism*. Durham, NC: Duke University Press, 2011.

Prashad, Vijay. *Everybody Was Kung Fu Fighting: Afro-Asian Connections and the Myth of Cultural Purity*. Boston: Beacon Press, 2001.

Probyn-Rapsey, Fiona. 'Human Rights and Animal Ethics Research Network Reading Group' [Zoom Meeting], 23 February 2021.

Quinlivan, Davina. 'Material Hauntings: The Kinaesthesia of Sound in *Innocence* (Hadžihalilović, 2004)'. *Studies in French Cinema* 9, no. 3 (2009): 215–24.

Quinlivan, Davina. *The Place of Breath in Cinema*. Edinburgh: Edinburgh University Press, 2012.

Radigue, Eliane. 'The Mysterious Power of the Infinitesimal'. *Leonardo Music Journal* 19 (2009): 47–9.

Radigue, Eliane. 'Interview about the Feedback Works: Paris, April 5th, 2011'. Interview by Emmanuel Holterbach, in the liner notes for Eliane Radigue, *Feedback Works 1969–70* (2014) [audio recording], Italy: Alga Marghen.

Rakow, Lana, and Laura A Wackwitz. *Feminist Communication Theory: Selections in Context*. Thousand Oaks, CA; London; New Delhi: Sage Publications, 2004.

Ramakrishnan, Kavita, Kathleen O'Reilly and Jessica Budds. 'The Temporal Fragility of Infrastructure: Theorizing Decay, Maintenance, and Repair'. *EPE: Nature and Space* 14, no 3 (2020): 674–95.

Rancière, Jacques. *The Politics of Aesthetics: The Distribution of the Sensible*. 2000. Translated by Gabriel Rockhill. London and New York: Continuum, 2004.

Reich, Steve. 'Pendulum Music'. In *Writings on Music, 1965–2000*, edited by Paul Hillier, 31–2. New York: Oxford University Press, 2002.

Reich, Steve. 'Interview with Thom Holmes'. *OHM: The Early Gurus of Electronic Music*, www.furious.com/perfect/ohm/reich.html, accessed 12 February 2021.

Republički zavod za statistiku. 'Prvi put registrovana drumska motorna i priključna vozila i saobraćajne nezgode na putevima, III kvartal 2020'. 2020, https://www.stat.gov.rs/sr-latn/vesti/20201123-prvi-put-registrovana -drumska- motorna-i-prikljucna-vozila-i-saobracajne-nezgode-na-putevim a-iii-kvartal- 2020/?s=1502, accessed 8 February 2021.

Rich, Adrienne. *On Lies, Secrets, and Silence*. New York: W. W. Norton & Company, 1979.

Richard, Frances. 'Music I've Seen: Christian Marclay'. *Aperture* 212 (2013): 25–33.

Riedweg, Christoph. *Pythagoras: His Life, Teaching, and Influence*. 2002. Translated by Steven Rendall in collaboration with Christoph Riedweg and Andreas Schatzmann. Ithaca: Cornell University Press; Bristol: University Presses Marketing, 2005.

Roberts, Mere and Peter R. Wills. 'Understanding Māori Epistemology: A Scientific Perspective'. In *Tribal Epistemologies: Essays in the Philosophy of Anthropology*, edited by Helmut Wautischer, 43–78. Aldershot and Brookfield: Ashgate, 1998.

Rockhill, Gabriel. 'Translator's Introduction: Jacques Rancière's Politics of Perception'. In *The Politics of Aesthetics: The Distribution of the Sensible*, edited by Jacques Rancière. 2000. Translated by Gabriel Rockhill. London and New York: Continuum, 2004.

Roh, David S., Betsy Huang, and Greta A. Niu. 'Technologizing Orientalism: An Introduction'. In *Techno-Orientalism: Imagining Asia in Speculative Fiction, History, and Media*, edited by David S. Roh, Betsy Huang, and Greta A. Niu, 1–19. New Brunswick: Rutgers University Press, 2015.

Rose, Deborah Bird and Thom Van Dooren. 'Guest Editors' Introduction'. In *Unloved Others: Death of the Disregarded in the Time of Extinctions*, edited by Deborah Bird Rose and Thom Van Dooren. *Australian Humanities Review* 50 (May 2011): 1–3.

Rose, Deborah Bird. 'Shimmer: When Everything You Love is Being Trashed'. In *Arts of Living on a Damaged Planet: Ghosts and Monsters of the Anthropocene*, edited by Anna Lowenhaupt Tsing, Heather Anne Swanson, Elaine Gan, and Nils Bubandt, 51–63. Minneapolis: University of Minnesota Press, 2017.

Rovelli, Carlo, *The Order of Time*. London: Penguin Books, 2017.

Russolo, Luigi. *The Art of Noise (Futurist Manifesto, 1913)*. Translated by Robert Filliou. New York: Something Else Press, 1967.

Sawada, Atsuko. 'Shingon shōmyō no kudensho ni tsuite 2: *Shōjitsushō*'. *Osaka kyoiku daigaku kiyo* 35, no. 2 (December 1986): 231–43.

Schaeffer, Pierre. *Treatise on Music Objects: An Essay Across Disciplines*. Translated by Christine North and John Dack. 1966. Berkeley: University of California Press, 2017.

Schafer, R. Murray. *The Soundscape: The Tuning of the World*. Rochester, VT: Destiny Books, 1993.

Schefer, Jean-Louis. *L'homme ordinaire du cinéma*. Paris: Cahiers du cinéma/ Gallimard, 1980.

Schmidt, Leigh Eric. *Hearing Things: Religion, Illusion, and the American Enlightenment*. Cambridge: Harvard University Press, 2000.

Schneider, Rebecca. *The Explicit Body in Performance*. London and New York: Routledge,1997.

Schor, Elana. 'McCain Sends out 'Truth Squad' to Counter Whisper Campaign'. *The Guardian*, 17 January 2008, https://www.theguardian.com/world/2008/ jan/17/usa.uselections20081, accessed 6 May 2020.

Schust, M. 'Effects of Low Frequency Noise up to 100 Hz'. *Noise and Health* 6, no. 23 (4 January 2004): 73.

Scott, Derek B. 'Orientalism and Musical Style'. *The Musical Quarterly* 82, no. 2 (1998): 309–35.

Scott, Kim. *Taboo*. Sydney: Pan Macmillan, 2017.

Seigworth, Gregory J. and Melissa Gregg. 'An Inventory of Shimmers'. In *The Affect Theory Reader*, edited by Melissa Gregg and Gregory Seigworth, 1–25. Durham, NC: Duke University Press, 2010.

Sharf, Robert H. *Coming to Terms with Chinese Buddhism: A Reading of the Treasure Store Treatise.* Honolulu: University of Hawai'i Press, 2002.

Shaviro, Steven. *The Cinematic Body: Theory Out of Bounds 2.* Minneapolis and London: University of Minnesota Press, 1993.

Shaviro, Steven. 'Untimely Bodies: Towards a Comparative Film Theory of Human Figures, Temporalities and Visibilities'. SCMS Conference, 2008, http://ftp.shaviro.com/Othertexts/SCMS08Response.pdf, accessed 27 November 2020.

Shaviro, Steven. *Post-Cinematic Affect.* Winchester: Zero Books, 2010.

Shearer, Rachel. 'Te Oro o te Ao: The Resounding of the World'. PhD diss., Auckland University of Technology, Auckland, New Zealand, 2018.

Sheppard, W. Anthony. 'An Exotic Enemy: Anti-Japanese Musical Propaganda in World War II Hollywood'. *Journal of the American Musicological Society* 54, no. 2 (2001): 303–57.

Sheppard, W. Anthony. *Extreme Exoticism: Japan in the American Musical Imagination.* Oxford: Oxford University Press, 2019.

Sholis, Brian. 'Shuffle'. *Print* 61, no. 4 (2007): 101.

Showalter, Elaine. *The Female Malady: Women, Madness, and English Culture, 1830–1980.* New York; London; Victoria; Ontario; Auckland: Penguin,1987; Pantheon Books, 1985.

'Shuffle'. *Lexico*, https://www.lexico.com/definition/shuffle, accessed 19 August 2021.

Silverman, Kaja. *The Acoustic Mirror: The Female Voice in Psychoanalysis and Cinema.* Bloomington and Indianapolis: Indiana University Press, 1988.

Silverton, Peter. 'Siouxsie & the Banshees: The Scream'. *Sounds.* 14 October 1978, https://www.rocksbackpages.com/Library/Article/siouxsie--the-banshees -the-scream, accessed 28 March 2022.

Sim, Stuart. *Manifesto for Silence: Confronting the Politics and Culture of Noise.* Edinburgh: Edinburgh University Press, 2007.

'Sing'. *Lexico*, https://www.lexico.com/definition/sing, accessed 9 March 2022.

Smith, Mick. 'Dis(appearance): Earth, Ethics and Apparently (In)Significant Others'. In *Unloved Others: Death of the Disregarded in the Time of Extinctions*, edited by Deborah Bird Rose and Thom Van Dooren. *Australian Humanities Review* 50 (May 2011): 23–43.

Smoller, Laura A. 'Playing Cards and Popular Culture in Sixteenth-Century Nuremberg'. *The Sixteenth Century Journal* 17, no. 2 (1986): 183–214.

Sobchack, Vivian. 'Waking Life: Vivian Sobchack on the Experience of Innocence'. *Film Comment* 41, no. 6 (2005): 46–9.

Solomon, Thomas. 'Music and Race in American Cartoons: Multimedia, Subject Position, and the Racial Imagination'. In *Music and Minorities from Around the World: Research, Documentation and Interdisciplinary Study*, edited by Ursula Hemetek, Essica Marks, and Adelaida Reyes, 142–66. Newcastle: Cambridge Scholars Publishing, 2014.

Stagoll, Cliff. 'Force'. In *The Deleuze Dictionary*, edited by Adrian Parr, 110–12. Rev. edn. Edinburgh: Edinburgh University Press, 2010.

Stalling, Jonathan. *Poetics of Emptiness: Transformations of Asian Thought in American Poetry*. New York: Fordham University Press, 2010.

Stewart, Kathleen. *Ordinary Affects*. Durham, NC: Duke University Press, 2007.

Stoever, Jennifer Lynn. *The Sonic Color Line: Race and the Cultural Politics of Listening*. New York: New York University Press, 2016.

Straebel, Volker. 'From Reproduction to Performance: Media-Specific Music for Compact Disc'. *Leonardo Music Journal* 19 (2009): 23–30.

Street, Sean. *Sound inside the Silence: Travels in the Sonic Imagination*. Cham: Palgrave Macmillan, 2020.

Stuart, Caleb. 'Damaged Sound: Glitching and Skipping Compact Discs in the Audio of Yasunao Tone, Nicolas Collins and Oval'. *Leonardo Music Journal* 13 (2003): 47–52.

Takahashi, Yukio. 'A Study on the Contribution of Body Vibrations to the Vibratory Sensation Induced by High-Level, Complex Low-Frequency Noise'. *Noise and Health* 13, no. 50 (1 January 2011): 2.

Takahashi, Yukio, Yoshiharu Yonekawa, Kazuo Kanada, and Setsuo Maeda. 'A Pilot Study on the Human Body Vibration Induced by Low Frequency Noise'. *Industrial Health* 37, no. 1 (1999): 28–35.

Takakusu, Junjirō, Watanabe Kaigyoku, and Ono Gemmyō, eds. *Taishō shinshū daizōkyō*. Vol. 77. Tokyo: Taishō Issaikyō Kankōkai, 1924–34.

Tallman, Susan. 'To The Last Syllable of Recorded Time: Christian Marclay'. *Art in Print* 6, no. 4 (2016): 10–15.

Taylor, Nicholas. 'Already Monstrous'. PhD Diss., Southern Cross University, Australia, 2020.

Taylor, Timothy D. *Global Pop: World Music, World Markets*. New York: Routledge, 1997.

Thompson, Marie. 'Whiteness and the Ontological Turn in Sound Studies'. *Parallax* 23, no. 3 (2017): 266–82.

Thompson, Marie. *Beyond Unwanted Sound: Noise, Affect and Aesthetic Moralism*. New York; London; Oxford; New Delhi; Sydney: Bloomsbury Academic, 2017.

Thompson, Marie, and Ian D. Biddle. *Sound, Music, Affect: Theorizing Sonic Experience*. New York: Bloomsbury Academic, 2013.

Tongson, Karen. 'China Girl'. In *Black Star Rising and the Purple Reign*, edited by Daphne Brooks. Durham, NC: Duke University Press, Forthcoming.

Tromans, Steve. 'Improvising Musical Experience: The Eternal Ex-temporization of Music Made Live'. In *Experiencing Liveness in Contemporary Performance: Interdisciplinary Perspectives*, edited by Matthew Reason and Anja Molle Lindelof, 178–87. New York and London: Routledge, 2016.

Tsou, Judy. 'Gendering Race: Stereotypes of Chinese Americans in Popular Sheet Music'. *Repercussions* 6, no. 2 (1997): 25–62.

van Eck, Cathy. *Between Air and Electricity: Microphones and Loudspeakers as Instruments*. New York: Bloomsbury Academic, 2018.

Varela, Francisco J. 'Whither Perceptual Meaning?' In *Understanding Origins: Contemporary Views on the Origin of Life, Mind, and Society*, edited by Francisco J. Varela and Jean-Pierre Dupuy, 235–63. Dordrecht: Kluwer Academic Publishers, 1992.

Wagatsuma, Ryūsho. 'Sandaihakkenki ni kansuru ikkōsatsu: Sōchō hon'i go'on sanjū kenritsu monsetsu o chūshin ni'. *Chisan kanga Kūkai* 64 (2015): 55–71.

Wald, Christina. *Hysteria Trauma and Melancholia: Performative Maladies in Anglophone Drama*. Basingstoke and New York: Palgrave MacMillan, 2007.

Waldrep, Shelton. 'The 'China Girl' Problem: Reconsidering David Bowie in the 1980s'. In *David Bowie: Critical Perspectives*, edited by Eoin Devereux, Aileen Dillane, and Martin J. Power, 147–59. New York: Routledge, 2015.

Walser, Robert. *Running with the Devil: Power, Gender, and Madness in Heavy Metal Music*. Hanover: University Press of New England, 1993.

Weiner, E. S. C., and J. A. Simpson. *The Oxford English Dictionary*, Vol. 18. 2nd ed. Oxford, UK: Clarendon Press, 1989.

West, David. *Opening Ceremony of the Games of the XXVII Olympiad in Sydney 15 September 2000 – Media Guide*. Lausanne: International Olympic Committee, 2000.

Williams, Linda. *Hardcore: Power, Pleasure and the Frenzy of the Visible*. Berkeley and Los Angeles: University of California Press, 1989.

Williams, Linda. 'Film Bodies: Gender, Genre and Excess'. *Film Quarterly* 44, no. 4 (Summer 1991): 2–13.

Williams, William Carlos. *Paterson*. Edited by Christopher MacGowan. New York: New Directions Publishing, 1995.

Wright, Herbert. 'Belgrade Waterfront: An Unlikely Place for Gulf Petrodollars to Settle'. 2015, https://www.theguardian.com/cities/2015/dec/10/belgrade-waterfront-gulf-petrodollars-exclusive-waterside-development, accessed 8 February 2021.

Yep, A. Gust, and B. Susan Shimanoff. 'The US Day of Silence: Sexualities, Silences, and the Will to Unsay in the Age of Empire'. In *Silence, Feminism, Power: Reflections at the Edges of Sound*, edited by Aimee Carrillo Rowe and Sheena Malhotra, 139–56. New York: Palgrave Macmillan, 2013.

Young, Eugene B., Gary Genosko, and Janell Watson. *The Deleuze and Guattari Dictionary*. London: Bloomsbury Academic, 2013.

Zeitlin, Judith T. 'From the Natural to the Instrumental: Chinese Theories of the Sounding Voice before the Modern Era'. In *The Voice as Something More: Essays towards Materiality*, edited by Martha Feldman and Judith T. Zeitlin, 54–76. Chicago: University of Chicago Press, 2019.

Zimmerman, Nadya. *Counterculture Kaleidoscope: Musical and Cultural Perspectives on Late Sixties San Francisco*. Ann Arbor: University of Michigan Press, 2008.

Žižek, Slavoj. *Violence*. New York: Picador, 2008.

Mediography

1984dreaming. 'James Blundell – Rain on a Tin Roof' *YouTube* (2014) [music video], https://youtu.be/ETPuJpfi5Pc, accessed 7 December 2021.

ABC RN, 'The Unsung Hero' (2019) [radio programme], https://vimeo.com /426488760, accessed 28 January 2022.

Aldrich, Robert. *Kiss Me Deadly* (1955) [film], USA: Parklane Pictures Inc.

Anderson, Brad. 'Sounds Like' [television episode], *Masters of Horror* (televised by Showtime on 17 November 2006), Canada and USA: Starz!/Nice Guy Productions/Industry Entertainment.

Aronofsky, Darren. *mother!* (2017) [film], USA: Paramount Pictures/Protozoa Pictures.

Bowie, David. 'China Girl' (1983) [audio recording], USA: EMI America.

Brötzmann, Peter. *Signs* (2002) [artist's multiple], Wuppertal: self-published.

Brötzmann, Peter. *Images* (2002) [artist's multiple], Wuppertal: self-published.

Cooper, Ian. 'Tin Symphony (Opening Ceremony) (Games of the XXVII Olympiad, 2000 Sydney Olympics) (Extended Version)' *YouTube* (2015) [music video], https://youtu.be/bTiUW7x0iMc, accessed 7 December 2021.

Day Above Ground, 'Asian Girlz' *YouTube* (2013) [music video], https://www .youtube.com/watch?v=qG1IfiuWuGY, accessed 30 March 2022.

Dickson, William K. L., and William Heise. *Blacksmith Scene* (1893) [film], USA: Edison Manufacturing Co.

Douglas, Carl. 'Kung Fu Fighting' (1974) [audio recording], USA: 20th Century Fox.

Griffith, D. W. *The Sealed Room* (1909) [film], USA: Biograph Company.

Griffith, D. W. *A Woman Scorned* (1911) [film], USA: Biograph Company.

Griffith, D. W. *An Unseen Enemy* (1911) [film], USA: Biograph Company.

Griffith, D. W. *Broken Blossoms or The Yellow Man and the Girl* (1919) [film], USA: D. W. Griffith Productions.

Hadžihalilović, Lucile. *Innocence* (2004) [film], Belgium, France, UK, and Japan: Ex Nihilo/Ateliers de Baere/Blue Light.

Hitchcock, Alfred. *Psycho* (1960) [film], USA: Shamley Productions.

International Olympic Committee. 'Sydney 2000 Opening Ceremony – Full Length' *YouTube* (2020) [film], https://youtu.be/qsLLzL27hYA, accessed 7 December 2021.

Klein, Eve. *Vocal Womb* (2018) [biomedical performance and installation], https://www.eveklein.com/vocal-womb, accessed 30 March 2022.

Low, *Double Negative* (2018) [audio recording], USA: Sub Pop.

Lucier, Alvin. *I am Sitting in a Room* (1993) [audio recording], USA: Lovely Music Ltd.

Lucier, Alvin. 'Alvin Lucier on a Lifetime of Experiments' interviewer unlisted. Red Bull Music Academy, *YouTube* (uploaded 31 May 2017) [video], https://www.youtube.com/watch?v=v-Pnb_ZE7Hs, accessed 8 February 2021.

Marclay, Christian. *Shuffle* (2007) [artist's multiple], New York: Aperture Foundation.

Martel, Lucrecia. *The Swamp* (2001) [film], Argentina: Wanda Visión, S.A.

Martel, Lucrecia. *The Headless Woman* (2008) [film], Argentina: Focus Features.

McKee, Maria, Eric Rackin and Jay Rifkin. 'Show Me Heaven' (1990) [audio recording], UK: Epic.

Miranda, Maria, and Norie Neumark. *It's Still Life: Listening to Compost* [installation], Melbourne, Australia: Living Museum of the West, 2019.

Muller, Charles A. '塵'. *Digital Dictionary of Buddhism*, http://www.buddhism-dict.net/cgi-bin/xpr-ddb.pl?q=塵, accessed 4 September 2020.

Nakadate, Laurel. *365 Days: A Catalogue of Tears* (2011) [photographic series].

Neumark, Norie. *Separation Anxiety: Not the Truth About Alchemy* (1997) [radiophonic work], http://www.somewhere.org/, accessed 28 January 2022.

Neumark, Norie. *Working with Worms* (2017–18) [blog], https://workingworms.net/, accessed 28 January 2022.

Neumark, Norie, and Maria Miranda. *Waiting* [online video component of art installation], *MoreArt, Melbourne, Australia*, 2017, https://vimeo.com/247735081, accessed 28 January 2022.

Noé, Gaspar. *Irréversible* (2002) [film], France: StudioCanal.

Nolan, Katherine. *Fight the Rising Odds* (2015) [performance to camera, digital video].

Nolan, Katherine. *Breathless* (2017) [performance to camera, digital video].

Nolan, Sidney. 'Ned Kelly' [enamel paint on composition board], Canberra: National Gallery of Australia, 1946, https://searchthecollection.nga.gov.au/object?uniqueId=28926, accessed 7 December 2021.

Oliveros, Pauline. *Sonic Meditations* (1971) [musical score], Urbana, IL: Smith Publications, 1974.

O'Reilly, Meara. 'Chaldni Singing'. *Artist's Statement*, https://mearaoreilly.com/Chladni-Singing, accessed 26 February 2021.

Paterson, Katie. *Sweet Country* (2008) [installation], https://katiepaterson.org/artwork/earth-moon-earth/, accessed 21 March 2022.

Pope.L. *Whispering Campaign* [installation], documenta 14, Kassel, Germany and Athens, Greece, 2017.

Pope.L. *Skin Set Drawings* [mixed media, selection], documenta 14, Kassel, Germany, 2017.

Radigue, Eliane. *Feedback Works 1969–70* (2014) [audio recording], Italy: Alga Marghen.

Radigue, Eliane. *Opus 17* (2015) [audio recording], Italy: Alga Marghen.

Radigue, Eliane. *Vice Versa, etc . . .* (2009) [audio recording], USA: Important Records.

Radigue, Eliane. 'Interview with Eliane Radigue' [video], interview by Evelyne Gayou. Translated by David Vaughn, *YouTube* (recorded 2013, uploaded

12 November 2018), https://www.youtube.com/watch?v=dByqwi7Jvbo&t
=1088s, accessed 6 February 2021.

River, Dylan, and Warwick Thornton. 'Sweet Country (Special Features: Dylan
River and Warwick Thornton)' (2018) [interview], Australia: Universal Sony
Pictures.

Robson, Mark. *Earthquake* (1974) [film], USA: Universal Pictures.

Rush, 'A Passage to Bangkok' (1976) [audio recording], Canada: Anthem
Records.

Simonetti, Gianni-Emilio. *Partita – Between Noise and Silence* (2013) [artist's
multiple], Molvena: Fondazione Bonotto and Flaneur&Dust.

Siouxsie and the Banshees, 'Hong Kong Garden' (1978) [audio recording], UK:
Polydor Records.

Solet, Paul. *Grace* (2009) [film], USA and Canada: ArieScope Pictures/Dark Eye
Entertainment/Leomax Entertainment.

Steinman, Jim, and Dean Pitchford. 'Holding out for a Hero' (1984) [audio
recording], UK: CBS Records.

Tait, Charles. *The Story of the Kelly Gang* (1906) [film], Canberra: National
Film and Sound Archive (2021), https://aso.gov.au/titles/features/story-kelly
-gang/, accessed 7 December 2021.

Thornton, Warwick. *Sweet Country* (2018) [film], Australia: Universal Sony
Pictures.

'Tinnitus Cure Frequencies 100Hz–23000Hz, Optimize Damaged Hearing'
YouTube (2021) [video], https://www.youtube.com/watch?v=tVLsivMpQCQ,
accessed 30 March 2022.

Tudor, David. *Pulsers/Untitled* (1984) [audio recording], USA: Lovely Music Ltd.

West, Kanye. *My Beautiful Dark Twisted Fantasy* (2010) [audio recording], USA:
Roc-a- Fella Records/Def Jam Recordings.

Index

Plate 1 Christian Marclay, *Shuffle* (2007), 75 playing cards, 74 four-colour images; box: 17.5 × 13 × 3.2 centimetres; cards: 16.8 × 12.1 centimetres; published by Aperture. Photo © Christian Marclay. Courtesy Paula Cooper Gallery, New York.

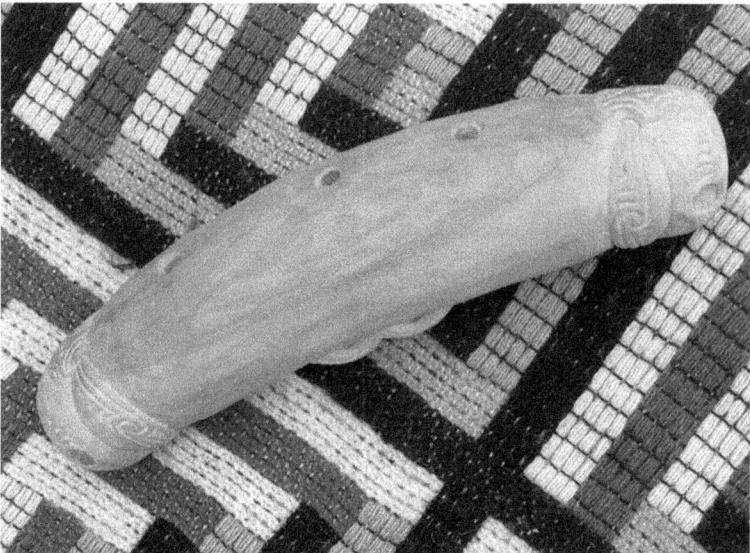

Plate 2 Ngōiro, the *kōauau* carved by Tim Codyre (2020). Photo credit: Rachel Shearer.

Plate 3 Video stills from *Breathless* (2017). © Katherine Nolan, digital video, with sound, 3 minutes 53 seconds.

Plate 4 A scene from the 'Tin Symphony' performance during the Opening Ceremony of the Sydney 2000 Olympic Games. Ned Kelly figures with guns ablaze march among corrugated iron sheets and tanks. Photo credit: Toby Melville. Courtesy Alamy.

Plate 5 Still from Warwick Thornton's *Sweet Country* (2018) with Sam Kelly (Hamilton Morris) chained, sitting in the outdoor court, with a tin shed in the background.

Plate 6 A medium shot frames an empty dirt road in Lucrecia Martel's *La Mujer Sin Cabeza* (2008).

Plate 7 A medium shot shows Vero and Vero's husband driving in a car in the dark night in Lucrecia Martel's *La Mujer Sin Cabeza* (2008). © Argentina Video Home 2008. All rights reserved.

Plate 8 A close-up shot shows Vero's face looking away from a backlit, blurry figure placed in the background of the shot in Lucrecia Martel's *La Mujer Sin Cabeza* (2008). © Argentina Video Home 2008. All rights reserved.

Plate 9 A close-up shot shows Vero's head in the foreground and a blurry silhouette in the background in Lucrecia Martel's *La Mujer Sin Cabeza* (2008). © Argentina Video Home 2008. All rights reserved.

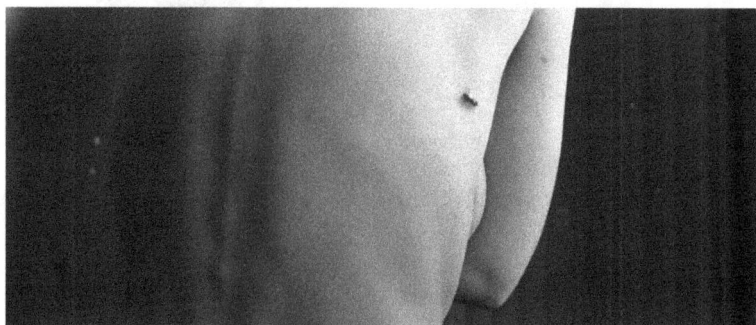

Plate 10 Madeline Matheson (Jordan Ladd) breastfeeding in the RV in Paul Solet's *Grace* (2009). © Anchor Bay Entertainment/Bremen Productions, LLC 2008. All rights reserved.

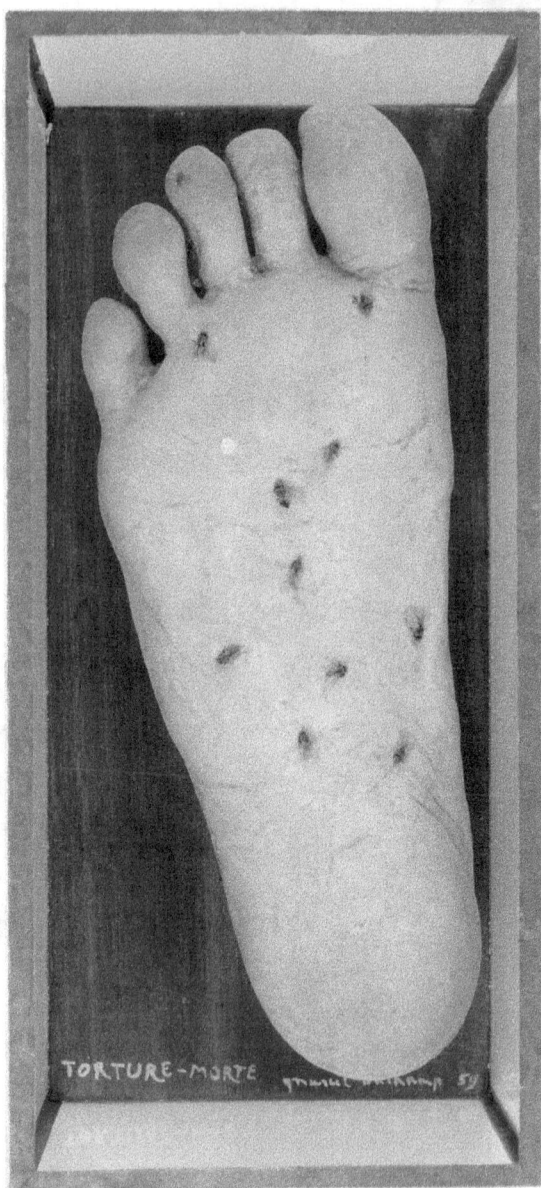

Plate 11 The torture of a fly-covered foot in Marcel Duchamp's *Torture-morte* (1959). © Marcel Duchamp. Societe Des Auteurs Dans Le Arts Graphiques Et Plastiques [ADAGP]/Copyright Agency, 2021. Photo credit: © Centre Pompidou, MNAM-CCI, Dist. RMN-Grand Palais/Jacques Faujour.

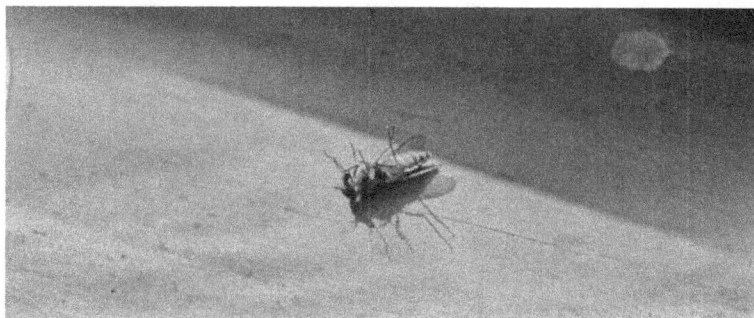

Plate 12 A fly dies at the windowsill in Darren Aronofsky's *mother!* (2017).

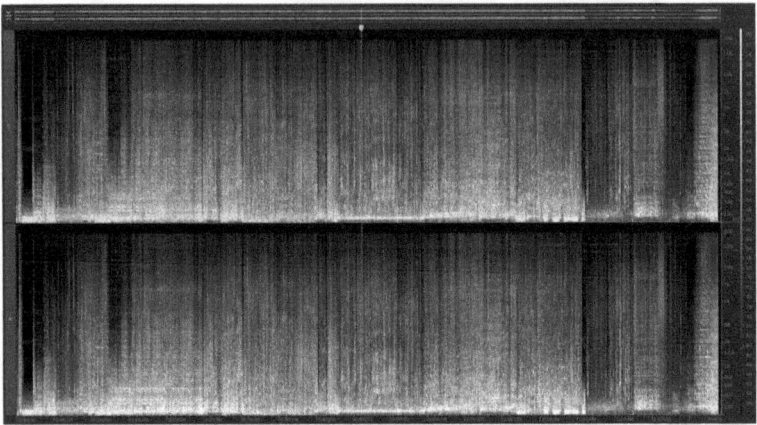

Plate 13 Spectrogram analysis frequency content of the whole film of Gaspar Noé's *Irréversible* (2002). This image shows both left (on top) and right (on bottom) channels. The horizontal axis represents time, and the vertical axis represents frequency from 20 Hz to 20 kHz bottom to top. The brighter areas are regions within the frequency spectrum with the most energy. Significant sub-bass content can be observed throughout the film, with peaks during the opening scenes and the rape scene. Spectrogram analysis by the author.

www.ingramcontent.com/pod-product-compliance
Lightning Source LLC
Chambersburg PA
CBHW060150280326
41932CB00012B/1702